Resource and Environmental Sciences Series

General Editors:
Sir Alan Cottrell, FRS
Professor T.R.E. Southwood, FRS

Already published:
Environmental Economics – Sir Alan Cottrell
Energy Resources – J. T. McMullan, R. Morgan and
R. B. Murray

Food, Energy and Society – David and Marcia Pimentel

Environmental Biology

E. J. W. Barrington, M.A., D.Sc., F.R.S.

Edward Arnold

© E. J. W. Barrington 1980

First published 1980 by
Edward Arnold (Publishers) Ltd
41 Bedford Square, London WCIB 3DQ

British Library Cataloguing in Publication Data

Barrington, Ernest James William
  Environmental biology. – (Resource and
  environmental sciences series).
  1. Biology
  I. Title   II. Series
  574   QH307.2

  ISBN 0–7131–2788–0

Photo typeset in India by
The Macmillan Co. of India Ltd.
Bangalore – 560 001

Printed in Great Britain by
Butler & Tanner Ltd, Frome and London

Preface

"Man is amazing, but he is not a masterpiece," he said . . ." Sometimes it seems to me that man is come where he is not wanted, where there is no place for him; for if not, why should he want all the place? Why should he run about here and there making a great noise about himself, talking about the stars, disturbing the blades of grass? . . ."

Joseph Conrad: Lord Jim.

The brief handed down to me was to present an account of the fundamentals of cell and organ biology for those students of the Resource and Environmental Sciences who do not have A-level biology. It was accompanied by the assurance that 'it would not really take a great deal of time'. This, however, proved over-optimistic, for, although my publishers have been generous in the matter of length, the field demanding my attention was so extensive that I have had to select, and prune, and prune again, all with great rigour. This was the more so because I wished not only to present the principles which govern the relationships of living organisms with each other and with their physical environment, but also to illustrate them in ways that would prove interesting to readers and would lead them to explore the subject further in the fascinating literature that is available for them. Definitions of the terms that they will need to master are incorporated into the text, and can be located by reference to the index.

It is a pleasure to acknowledge my indebtedness to Professor T. R. E. Southwood, F.R.S. for inviting me to write this book, and then for reading the typescript with a close attention that brought many improvements to light, and to the staff of Edward Arnold, for giving me the benefit of their patience and skill. My thanks are due also to the publishers and authors who have allowed me to use copyright material in certain of the text-figures, the sources of which are given below each one.

Alderton,   E. J. W. Barrington
Tewkesbury, Gloucestershire
1980

Contents

1 Life and the Cell

What is Life?

Biology is the study of living organisms – but it is easier to say this than to define exactly what we mean by life, although we feel that we can recognize it when we see it. Unfortunately, we only see it in the form that it takes within the biosphere of this planet. We do not know what it might look like elsewhere in the universe, and it has been suggested that it is this limited perspective which makes definition of it so difficult. Perhaps so; but we must take life as we find it, and seek for definitions within that limit.

Fundamentally important is the close integration of life with the environment in which it is maintained; an integration so close that all biology might be described as environmental biology. Living organisms are characterized chemically by their ability to maintain a continuous dynamic relationship with their physical or non-living (abiotic) environment, and with each other (their biotic environment), exchanging materials and carrying out delicately controlled manipulations of these in a network of integrated chemical processes that constitute metabolism. These metabolic reactions take place in an aqueous complex called protoplasm. So it is that the presence of water, which can constitute 70–90 % or more of living material, is an essential condition for the maintenance of life.

Metabolism is concerned in part with anabolism and in part with catabolism. Anabolism involves the building up of giant molecules containing stores of potential energy which determine the capacity of the system to do work. Catabolism involves the breaking down of these molecules, with the release of some of their stored energy, which thus becomes available as free energy. It is the release of this energy, which is called high quality energy because it can be expended in useful work (see Cottrell, 1978, in this series), that makes possible the many types of activity that are such a feature of life.

This analysis draws our attention to a fundamental difference between living systems and non-living ones. The latter do not exchange materials with their environment; we say that they are closed systems. Living ones, by contrast, are open systems, maintaining a dynamic steady state which depends upon continuous input and output of materials, together with conversion of the energy associated with these materials. But the handling of energy by any system, living or non-living, is governed by the First and Second Laws of Thermodynamics, and here life confronts us with what seems at first to be a paradox. The First Law states that the total energy of the universe remains constant, although it can be converted from one form to another. The Second Law states that the materials of the universe are tending towards a random or disorderly state, accompanied by the conversion of the potential energy of the more orderly material into the kinetic energy of random molecular movement. This form of kinetic energy, which is not available for work, and which is therefore low quality energy, is said to be rich in entropy; we may therefore say that the universe is tending towards a condition of maximum entropy, or of minimal free energy.

But if living systems are subject to these Laws, how can plants and animals build up stored potential energy, release free energy, and yet still maintain their form and organization? Clearly they must have some way of escaping, at least temporarily, from the trend to increasing entropy. The escape is made possible because they can use radiant energy from the sun and convert it into the potential energy of complex chemical compounds. Later, these are indeed broken down, with the release and expenditure of free energy, so that the escape of life from the Second Law of Thermodynamics can be only a temporary one, for there is a limit to the energy resources of the sun. In the long run we are all dead, the only uncertainty being the extent to which this movement towards the extinction of life may be accelerated by the activities of our own species.

But organisms do very much more than simply maintain themselves. They are capable also of growth, which enables them to develop into large and complex structures, based upon the replication of individual units of organization that we call cells (p. 10). Furthermore, they are also able to reproduce, generating offspring which go through the same processes of growth and replication. This is possible because offspring receive from their parents genetic material carrying units of coded information called genes. These establish the identifying characteristics of the offspring, and pass in due course to the next generation after a process of replication that we consider later.

Associated with this replication and reproduction is yet another characteristic of living organisms: their capacity to improve their relationship with their environment and thereby to become increasingly efficient in ensuring their own survival and the production of offspring to continue their line. This increasing efficiency in environmental relationships, which is one aspect of biological progress, depends upon a factor of variability which is built into living systems. In part this variability is a direct response to the environment, as when plants grow more luxuriantly in a rich soil than in a poor one. This type of variation is not inherited, and affects only the individual concerned, except in so far as offspring may well profit in a general way by having stronger parents. But variability also arises, in ways to be considered later, from heritable changes in the genetic material. This results in some organisms developing characteristics that give them an adaptive advantage in survival over their fellows. They tend, because of this, to leave more offspring than their less well adapted fellows. So, because this type of variation is inherited, they pass on the improvements to subsequent generations to an extent that brings about a shift in the characteristics of the population. This, expressed here in the simplest possible terms, is the basis of Darwinian evolution by natural selection. By favouring those that are 'fit', in the sense of being reproductively more effective, it results in cumulative change and diversification in characteristics.

One index of the reality of evolution is the existence today of primitive organisms, which are so called because they resemble earlier stages of evolutionary history. They contrast with those which represent later stages. These are said to be specialized, because they have developed many adaptations which have increased their efficiency but which tend to limit the possibilities of their further evolution. These two categories are not mutually exclusive; biological categorization is rarely as simple as that. Primitive organisms may be expected to have developed some specializations that are essential for their survival, while specialized ones will often have retained some primitive features. Used with judgement, however, the terms 'primitive' and 'specialized' are valuable, and are an essential and frequently encountered element in biological analysis.

These, then, are some of the characteristics of life to which we shall be giving detailed attention: metabolism, the conversion of energy, and the maintenance of dynamic steady states in open systems; growth, replication and reproduction; variation, evolution by natural selection, and improvement in the effectiveness of environmental relationships. These characteristics are expressed in a vast diversity of structural and functional detail, but the treatment of them here must be highly selective. We shall therefore concentrate upon those aspects of life which seem especially relevant to our understanding of the environment, both non-living and living, in which we seek to survive. For this reason, we shall concern ourselves more with functional aspects than with structural ones.

The Chemical Basis of Life

What are the materials which make life possible, and how are they put together? Particularly characteristic of them are a small number of types of giant molecules (macromolecules), composed of smaller molecules which are being continuously put together and taken apart. Some of these macromolecules are structural, responsible for the form and support of the body. Others are involved in energy conversion, while yet others constitute the inherited genetic material. But whatever their function, few remain for long unchanged. Not the least remarkable feature of biological open systems is the turnover of material which takes place even within such apparently inert structures as the skeleton of vertebrate animals, and which has been clearly demonstrated by following the movement of tracer quantities of radioactive isotopes into and out of the body.

Four main categories of macromolecules can be distinguished. All are compounds of carbon, dependent for their complexity and diversity upon the tetravalency of that element, and upon the capacity of carbon atoms to join with each other to form a variety of patterns of chains and rings.

One category comprises the carbohydrates, which are in part structural, and which also provide major reserves of energy. They have the empirical formula $(CH_2O)_n$, and are cyclic compounds, some of their carbon atoms forming continuous rings.

Carbohydrates with a single ring (monosaccharides) are exemplified by glucose, which is a hexose (n being 6). In its perspective formula (Fig. 1.1) the ring is considered to lie in a plane at right angles to the page, the heavy lines showing those parts of the ring lying above the plane of the paper, and the light lines those parts lying below that plane. Similarly, the hydroxyl groups lie above or below. Carbon atom 1 is an asymmetric centre, so that its attached hydroxyl group can be orientated in one or other of two ways. In consequence, glucose has two isomeric forms, termed respectively α-glucose and β-glucose.

Monosaccharides are examples of molecular units (monomers) that can combine to form larger molecules called polymers. Maltose (Fig. 1.1), for example, is a disaccharide (dimer), formed by the combination of two α-glucose units with the loss of a molecule of water; cellobiose is similarly formed from two β-glucose units. This mode of combination, involving the loss of water, is termed condensation. We shall refer later to the opposite process, termed hydrolysis, in which the polymers are broken down into their units with the addition of water. Starch and glycogen, which are major energy stores of plants and animals respectively, are large polymers, composed of thousands of α-glucose units; cellulose, of great

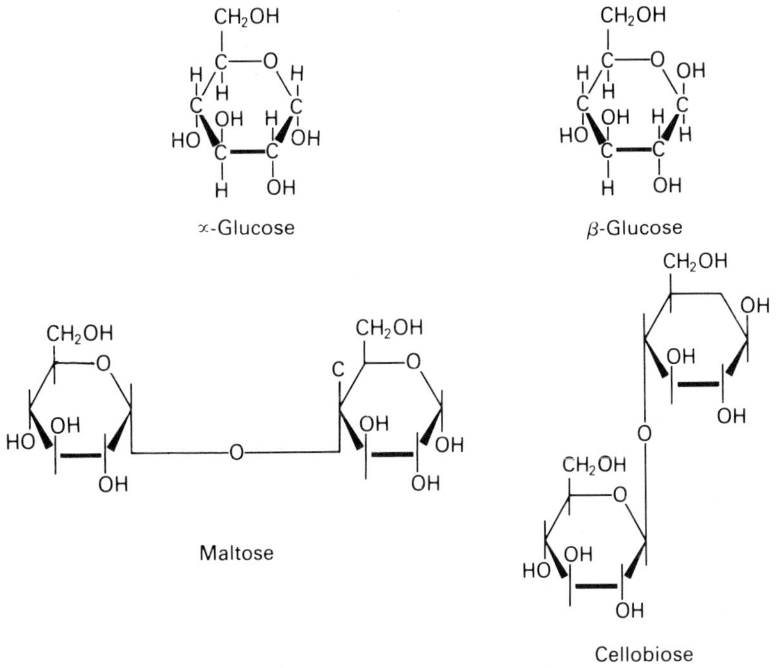

Fig. 1.1 Chemical structure of some carbohydrates. In starch the glucose molecules are joined by 1:4-α-glycosidic links as in the disaccharide maltose. In cellulose they are joined by 1:4-β-glycosidic links as in the disaccharide cellobiose.

structural importance in plants, is a comparably large polymer of β-glucose. These exemplify the macromolecules mentioned earlier.

Another group of macromolecules comprises the lipids, a varied assemblage of substances which share the common property of being insoluble in water but soluble in ether and other organic solvents. Lipids may be either simple lipids, which are esters of fatty acids, or compound lipids, which are derivatives of them. Simple lipids, which are water-repellent (hydrophobic), include fats (triglycerides, in which one molecule of glycerol is esterified with three of fatty acids, see Fig. 1.2), and waxes, which are esters formed from higher alcohols and higher fatty acids. Fats constitute important energy stores in both plants and animals, while waxes have functions in waterproofing and in other forms of protection. Compound lipids, which possess one or more additional groups with an affinity for water (hydrophilic), have important structural functions, resulting from their tendency to come together to form surface films and membranes.

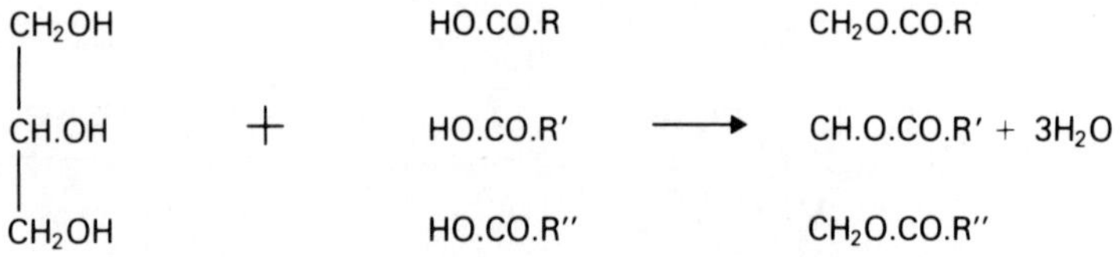

$$
\begin{array}{lll}
\text{CH}_2\text{OH} & \text{HO.CO.R} & \text{CH}_2\text{O.CO.R} \\[2mm]
| & & \\[1mm]
\text{CH.OH} \quad + & \text{HO.CO.R}' \longrightarrow & \text{CH.O.CO.R}' + 3\text{H}_2\text{O} \\[2mm]
| & & \\[1mm]
\text{CH}_2\text{OH} & \text{HO.CO.R}'' & \text{CH}_2\text{O.CO.R}''
\end{array}
$$

Glycerol	Fatty acid molecules	Triglyceride

Fig. 1.2 Formation of a triglyceride fat by the combination of glycerol with three fatty acid molecules. R, R′, R″, side chains of the fatty acid.

A third group of macromolecules comprises the proteins, which are of outstanding structural and metabolic importance, not least because they form the enzymes which promote and regulate metabolic processes. They are composed of carbon, hydrogen, oxygen and nitrogen, often with sulphur as well, and sometimes with phosphorus. Their structural units are amino acids (Fig. 1.3), those characteristic of living organisms being α-amino acids, with the amino group carried by the carbon atom that is adjacent to the carboxyl group. Proteins are built up by the condensation of a carboxyl group of one amino acid with an amino group of another, the amide (–CO.NH–) link between them being termed a peptide bond (Fig. 1.3). Compounds formed in this way are called peptides (dipeptides or tripeptides, for example, or polypeptides when many

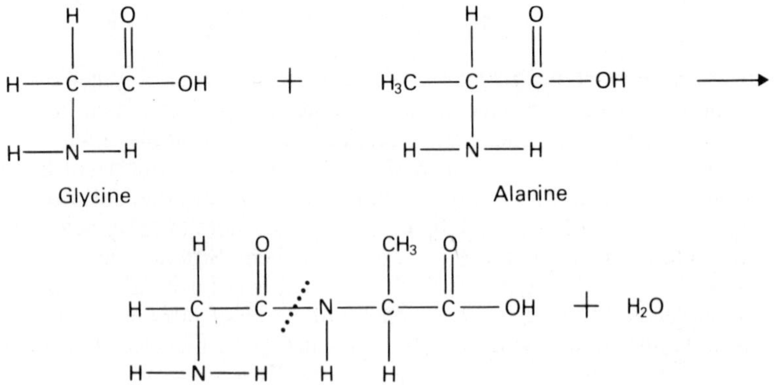

Glycine   Alanine

Alanylglycine

Fig. 1.3 Formation of a dipeptide by the combination of two amino acids. The dotted line indicates the peptide bond.

amino acids are involved, but with the molecule still below the size of a typical protein).

The fourth group of macromolecules comprises those which carry the coded information to which we have already referred. There will be much to say of these later. For the present, it will be sufficient to mention that they comprise two categories, deoxyribonucleic acids (DNA) and ribonucleic acids (RNA), all of them being polymers of nucleotides (Fig. 1.4). Each nucleotide of DNA consists of a molecule of phosphoric acid, a pentose sugar (deoxyribose), and either a purine base (adenine or guanine) or a pyrimidine base (cytosine or thymine). Each nucleotide of RNA is similarly constructed in principle, but with ribose instead of deoxyribose, and with uracil replacing thymine.

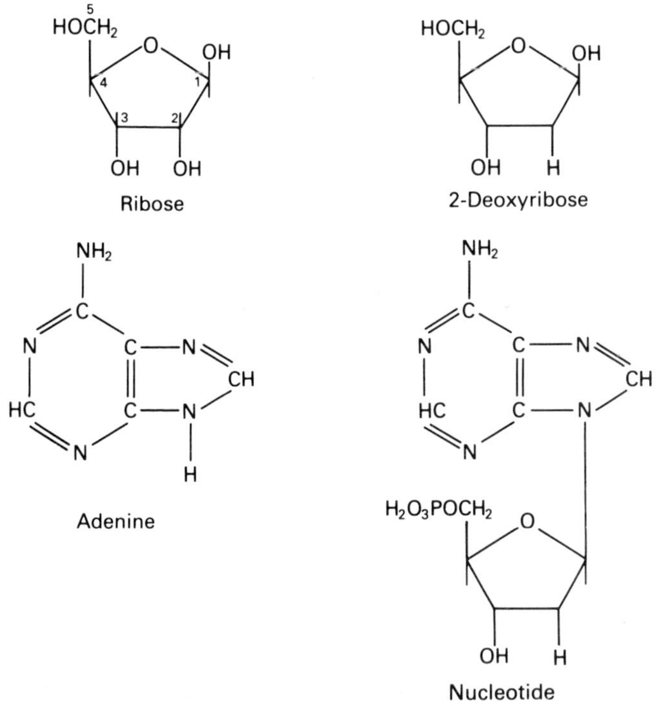

Fig. 1.4 Chemical structure of a nucleotide and its components.

In addition to these macromolecules, there are also small molecules, formed of a limited range of inorganic ions that are essential in one way or another for the maintenance of life.

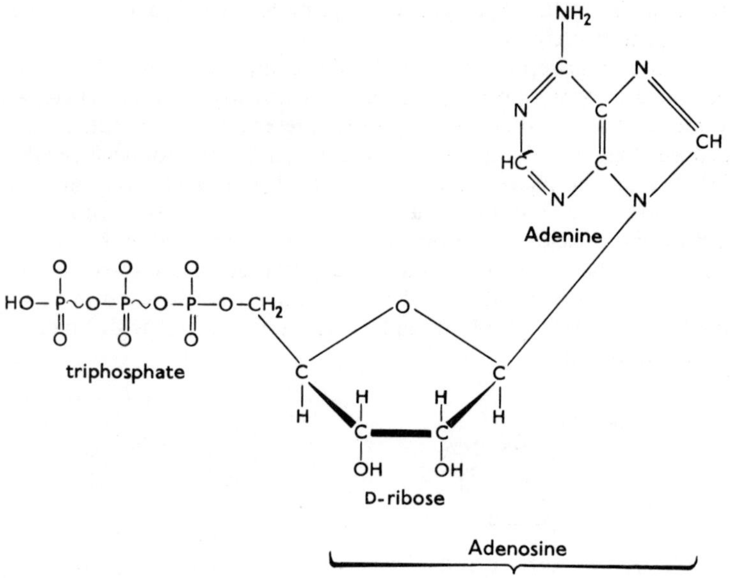

Fig. 1.5 Chemical structure of adenosine triphosphate (ATP).

Phosphate is of unique value because of the key role played by adenosine triphosphate (ATP) in energy conversion within the cell. It will be noted (Fig. 1.5) that one molecule of adenosine (a nucleoside formed by the combination of adenine with d-ribose) combines with three molecules of orthophosphoric acid to form ATP. The terminal phosphate group can be readily removed by hydrolysis with the release of a large amount of energy, equivalent to 30.56 kJ(= 7.3 kcal) and the formation of adenosine diphosphate (ADP). ATP is therefore referred to as an energy-rich or high energy compound, and the terminal phosphate bond as a high-energy bond, represented conventionally by the sign ~. However, this representation is only a matter of convenience, for the energy is not localized in a particular bond, but is a property of the molecule as a whole. The terminal phosphate bond of ADP can also be removed, with the formation of adenosine monophosphate (AMP), and with the release of the same amount of energy; its terminal bond is therefore represented in the same way. These phosphate compounds are of crucial metabolic importance, not only because they provide for the storage of potential energy, but because their enzymatically controlled hydrolysis provides for the release of this energy in small quantities that can be handled by living organisms without damaging themselves (p. 42).

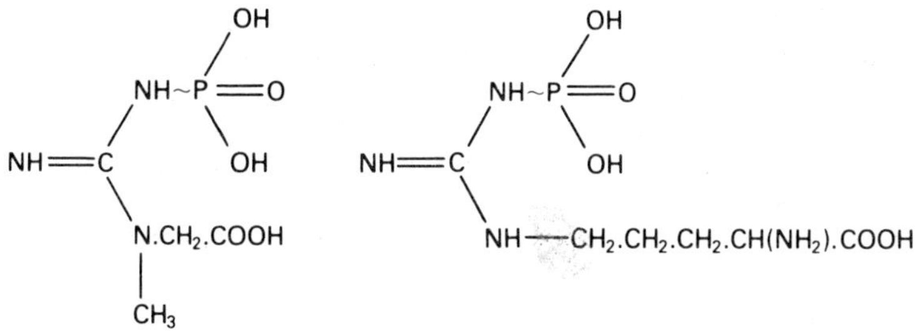

Creatine phosphate (phosphocreatine)   Arginine phosphate (phospho-arginine)

Fig. 1.6 Chemical structure of phosphagens.

Phosphates contribute also to other high-energy compounds, examples being the phosphagens (Fig. 1.6), in which the energy-rich bond is a phosphate-amide linkage. Creatine phosphate (phosphocreatine) is characteristic of vertebrates, while related compounds, including arginine phosphate (phospho-arginine), are found in invertebrates. They are important when ATP is used up more quickly than it can be regenerated from ADP, as can very easily happen in muscular tissue. A reversible reaction permits the phosphagen to release energy which can be made available at these times of emergency for the generation of ATP:

Creatine phosphate + ADP ⇌ Creatine + ATP

Following the release of energy, ADP is regenerated from AMP, and ATP from ADP, provided, of course, that there is an adequate input of energy into the cell to provide for this renewed storage.

Calcium, magnesium, sodium and potassium are examples of other ions of major importance. Enzymes often require to have associated with them additional components of low molecular weight. These are called co-enzymes, or, if they are too firmly bound to be separated by dialysis, prosthetic groups. They may be complex organic compounds which are often chemically changed during the reactions in which they are involved, being then restored to their original state by other enzymes. Chains of enzymes, of which we shall see examples later, may be built up in this way. Other co-enzymes may be metallic cations, such as iron, copper or magnesium; these are sometimes called activators, and the same term is applied to inorganic ions which increase the activity of enzymes without being essential for their functioning. Many other uses of ions could be listed. Thus calcium is of structural importance in both plants and animals;

phosphate contributes to structure in many animals, in addition to being universally important in organisms because of its role in energy conversion; magnesium is an essential component of the chlorophyll of plants; while iron is of comparable importance in the formation of the cytochromes of the electron transport chain (p. 40), and of the haemoglobin which transports oxygen in vertebrates and in some invertebrates. We refer later to some of these activities in more detail.

The Cell

A review of the chemical basis of life raises an obvious and fundamental question. How is this material, distributed in a watery solution and suspension, so structured and organized as to make possible the execution of complex and interlocking activities? The answer is found in the subdivision of the protoplasm into units called cells, within each of which it is typically differentiated into nucleus and cytoplasm.

Many plants and animals are formed of vast numbers of cells (perhaps up to 10^{11} in the human body), all of them derived by division of pre-existing cells. Such organisms are said to be multicellular. It is characteristic of them that division of labour occurs amongst the cells, particular cells being specialized to carry out particular functions. Equally characteristic, and a logical design consequence, is the association together of cells of like function to form tissues, and, more especially in animals, for complex functions to be carried out by a number of different types of tissue associated together to form organs. But there are many other organisms (the animals classified as Protozoa are an example), which are minute in size, and which have a structure equivalent to that of a single cell; they are sometimes termed unicellular, sometimes acellular (or noncellular). This, however, need not effect our analysis of the structure and functioning of the cell, in which is found the key to the organization of life.

That the key must be there was well understood in the nineteenth century, but the analysis was limited until recently by the restrictions of the light microscope, which permits magnifications of at best about 1500 times. The limit is imposed by the degree of resolution permitted by the wavelength of light. Points which are close together cannot be visually resolved as separate points when they are closer than about half the wavelength of the light which is illuminating them. With white light this distance is about 0.25 μm. Increase of magnification cannot influence this limit, and cannot, therefore, reveal further detail. Immense advances have been made possible, however, by the introduction of the electron microscope, in which electrons are used as the illuminant, for their very short wavelength permits a resolution of about 8–10 nm. It is thus possible to prepare visual records (electron micrographs) with magnifications of as much as 160 000 times, a technical achievement which has revealed ultrastructural detail that cannot be seen at all with the light microscope. (One μm is 10^{-6} m; one nm is 10^{-9} m.)

By itself, this would not be enough. With no further information than that of ultrastructure, we should be in the position of anatomists studying form divorced from function. What has revolutionized our grasp of cell organization is the conjunction of electron microscopy with developments of biochemical analysis that allow the separation and identification of exceedingly minute quantities of materials, and the use of radioactive isotopes to follow the fate of these materials in the cell. Further essential advances have been applications of chemical reactions to the study of microscope sections; these techniques of histochemistry and cytochemistry make it possible to identify specific materials within the nucleus and cytoplasm, and to determine their precise localization.

Plasma Membrane

Two requirements are fundamental for the functioning of the cell (Fig. 1.12, p. 24). First, it must be separated at its surface from the medium in which it lives, so that it can preserve its independent existence, yet this separation must permit controllable exchanges with the environment. Secondly, the cell must be organized in such a way that a diversity of activities can be carried on side by side, with or without interactions between them, and always within microscopically small areas. The daunting design problems need no emphasis.

The cells of both animals and plants are covered by a membrane (cell membrane, plasma membrane, plasmalemma) which is part of the living substance of the cell, and which, in plants, underlies the cell wall (see later). Localization and integration of cell functions depends upon the extensive use of membranes, not only over the surface but within the cell as well. There must be differences in detail between them, but the structure of the plasma membrane will serve to exemplify the general principles of their organization, and it is certainly the most thoroughly investigated.

Electron micrographs show it to consist of two electron-dense (dark) layers, each about 2.4 nm thick, with an electron-lucent (light) gap of the same thickness lying between them. There is some disagreement as to the precise explanation of this appearance, but the composition of the membrane can be stated in general terms with reasonable assurance (Fig. 1.7). Lipids are one important component, and proteins (mainly responsible for the electron-dense layers) are another, the latter having two main functions. In part they are structural, contributing to the elasticity of the membrane and to its stability, but in part they are enzymes, for the membrane is a metabolically active structure, while some are receptors for chemical signals (p. 198). Cholesterol is also present in the membrane, probably mainly as a structural component, and there is also some carbohydrate.

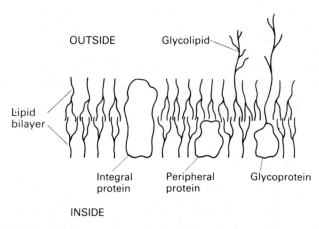

Fig. 1.7   A possible model of the structure of the plasma membrane. Much of it is a fluid bilayer. Globular proteins extend across the membrane, while peripheral proteins are mainly associated with the inner surface. Glycolipids and glycoproteins form a surface layer called the glycocalyx. (From Oschman *et al.* (1974). *Symp. Soc. exp. Biol.*, **28**, 305–50.)

The likely arrangement of the various components has to be inferred from both the ultrastructure and the functioning of the membrane. One property of major importance is its permeability, which is both selective and controllable. Water can usually pass freely through it, and fat-soluble substances do so as well, because of its lipid content. Water-soluble particles can also pass through, and are enabled to do so by more than one process. It is supposed that one factor permitting their passage is the presence of pores in the membrane. If these are of adequate size, neutral particles will tend to diffuse down their concentration gradients. Another factor is that there can be potential differences across the membrane, reaching values of up to 100 mV. This can cause charged particles to diffuse down the electrochemical gradient, and against the concentration gradient.

Movement of particles against both their concentration and

electrical gradients is also very common. This type of movement, called active transport, depends on processes which take place within the membrane and require the expenditure of energy. Organic materials such as monosaccharides and amino acids are transported in this way, and so also are a number of inorganic ions. Active transport is thought to be effected through the mediation of lipid-soluble carrier molecules which move to and fro across the membrane, taking up the transported particle into loose combination on one side and releasing it at the other side. The energy required for this is provided by ATP; active transport can therefore be interrupted by the use of metabolic inhibitors, such as potassium cyanide, which interfere with ATP metabolism, and which thus provide valuable experimental tools in the investigation of cell functions.

Two other methods by which materials may be transported across the plasma membrane must be briefly mentioned. One is exchange diffusion, in which an ion is taken up on one side of the membrane and exchanged with another ion of the same chemical species on the other side. This process, which is obscure, and perhaps of little importance, may also depend upon the action of carrier molecules, although this is uncertain.

Another method is pinocytosis (Fig. 1.12), in which the cell surface invaginates and engulfs material from the surrounding medium, taking it up into the cytoplasm in a vacuole. This process, which may be preceded by adsorption of particles to the plasma membrane, makes a significant contribution to the ingestion of food material in animals, either directly from the outside, as in *Amoeba*, or as part of the digestive activity of the alimentary epithelium of higher forms.

Given these facts of ultrastructure and function, how far can we explain the organization of the plasma membrane? One suggested model, derived from an earlier one formulated by Davson and Danielli, is shown in Fig. 1.7. According to this, the membrane is a mosaic, the central electron-lucent region being a bimolecular layer of phospholipid chains, while the dense outer and inner layers are composed of protein in association with the polar groups of the chains. Some of this material is combined with carbohydrate to form glycolipid and glycoprotein. Pores must be present, although they are too small to be visualized even by electron microscopy. Some would presumably be permanent, but it has been thought that thermal movement of the membrane molecules could provide temporary pores, or permit variation in the size of the permanent ones, thus giving one possible basis for the regulation of membrane permeability. Whether the proteins are confined to the inner and outer surfaces of the membrane, as postulated in the earlier model, is uncertain. It now seems more likely that protein is also present within the substance of the membrane, perhaps lining the pores, and perhaps associated quite closely with the phospholipid chains (Fig. 1.7). There is evidence, for example, that some of the enzymes of the membrane can only be fully functional if they are associated with some lipid material.

Considering the complexity of membrane activities, and the ultrastructural level at which they are operating, some divergence in the interpretations of membrane structure are only to be expected, but we can feel reasonably sure of the general principles outlined here. It is customary to refer to this type of membrane as a unit membrane. The same principles are thought to apply also to the various intracellular membranes which we shall be considering, although there is probably much diversity of detail in relation to diversity of function.

Cell Wall

A feature of plant organization which is not found in animals is the cell wall. The description of this by Robert Hooke, who observed it in sections of cork, was one of the earliest fruits of seventeenth-century microscopy. It is a non-living structure, which is formed externally to the cell, and which contributes to its protection and support. Some 50% of it consists of long-chain polymers of α-cellulose, arranged as microfibrils which provide the structural reinforcement of the wall. In addition, there is 5% pectin (an acid mucopolysaccharide), 25% hemicellulose (formed of sugars other than glucose) and 20% protein, these materials together forming a matrix which constrains extensibility.

The wall is first laid down by the cell as the primary cell wall, which, at least in dicotyledons (p. 48), may later undergo secondary thickening by deposition of polysaccharides (mainly cellulose) on its inner surface, while both primary and secondary walls may be impregnated with additional materials. Thus cells at plant surfaces (epidermal cells) are commonly protected by an impervious cuticle, formed by the deposition of waxy materials (cutin and suberin) in their cell walls. The composition of this cuticle is of some practical importance to man, for waxes increase resistance to wetting, and thus to the action of herbicides and fungicides.

Another addition, and one of great structural importance, is lignin. This is a substance of high molecular weight, derived from carbohydrate by a complex sequence of reactions involving the polymerization of phenolic alcohols. Lignification involves the formation of a matrix within which are enclosed cellulose microfibrils, the whole providing a material which has great tensile strength, derived from the microfibrils, and great rigidity, derived from the lignin. It is the latter which makes possible giant forest trees, for lignin constitutes up to 38 % of their wood.

Despite the firmness of the plant cell wall, the cells are not functionally isolated by it. Depressions in it are points of thinning which are called pores, although they seem normally to be closed by thin membranes. Across these, and elsewhere as well, extend strands of cytoplasm called plasmodesmata, so that the contents of one cell are connected with those of another.

The Nucleus

The DNA in the cells of most organisms (eukaryotes) is present in a highly organized nucleus, but in some primitive forms (prokaryotes, comprising bacteria and blue-green algae) the nuclear arrangement is simpler. In these there are no nuclear membranes (see below), nor is the DNA combined with protein; further, bacterial cells may have additional autonomous DNA elements called plasmids. Prokaryotes, which will be encountered later in other contexts, are presumably related to the green plants, and are sometimes classified with them in the plant kingdom, although it may be preferable to regard them as constituting a separate group or kingdom of their own (p. 230).

Standing apart from them, although sometimes thought of as prokaryotes, are the viruses. These are collections of minute particles called virions, 20–400 nm in size. Each virion particle consists of a core of nucleic acid (DNA or RNA) enclosed in a sheath of protein or of protein and lipid; some protein may also be present in the core. Virions are peculiar in that they are found only as infective agents within the cells of prokaryote, plant or animal hosts, their protein sheath being sometimes highly specialized to secure the injection of the core into the host cell. Within these cells they multiply by replication of their nucleic acid (p. 22), using their genome to direct the protein synthesizing machinery of the host in their own interest, so that it provides the protein needed to complete the assembly of the virion particles. The origin of viruses is uncertain, one possible interpretation being that they are groups of genes which became isolated within prokaryote or eukaryote cells and developed an individual but parasitic (p. 45) mode of life. An intriguing suggestion is that virions, aptly described as the most numerous genetic objects in the biosphere, might have played a significant part in evolution by transferring genes from one organism to another. What is certain is that some of them are major agents of disease in plants and animals, although many others are doubtless without overt effects and therefore remain undetected. They are thus of immense medical and agricultural importance, and are under investigation as a practicable means of controlling insect pests, more economical and more acceptable than pollution by chemical insecticides, but still with many hazards to be evaluated.

The separation of the nucleus from the cytoplasm in eukaryote cells is maintained by the enclosure of the former within two nuclear membranes, separated from each other by a space of about 14 nm, which is so narrow that the two membranes appear under light microscopy to be a single one. The ground substance of the nucleus (nucleoplasm) contains pairs of thread-like bodies (23 pairs in human cells) which are called chromosomes because they are readily stained by certain dyes. They consist of nucleoprotein, which is a complex of protein (especially histones) and of the DNA which carries the coded genetic information.

Within the nucleus there are usually one or more bodies called nucleoli, formed at particular points on one or other of the chromosomes, and consisting of RNA, unbounded by membranes. The number and appearance of these bodies depends upon the condition of the cell. Thus, a single large nucleolus is often seen in a cell engaged in vigorous protein synthesis, while nucleoli disappear at times when the cell is dividing. It seems that they are reserves of RNA, laid down under the influence of special regions of the chromosomes to which they are attached, and which are coded for their production. Full expression of the information carried by DNA requires it to continue influencing the activities of many of the cells throughout life. While, therefore, the independent organization of the nucleus and its chromosomes must be maintained, there must also be some path of communication between nucleus and cytoplasm, for it is within the cytoplasm that the coded instructions carried by the chromosomes are ultimately expressed.This communication is possible because the nuclear membranes contain pores, and also have extensions which form channels running into the cytoplasm and joining up with other membranous structures to be mentioned later. These pores and channels provide the pathways by which material can pass out of the nucleus, and into it as well, for the control of cell functions demands two-way communication between nucleus and cytoplasm.

The Genetic Code

The DNA molecule, which, as we have seen, is a nucleotide polymer, consists of two long chains or strands of alternating deoxyribose and phosphoric acid residues, each deoxyribose residue being linked with one of the four nitrogeneous bases (adenine, cytosine, guanine, or thymine). Each of the strands is coiled into a helix, the two being wound in opposite directions around a common centre in which are located the bases. These are held together in pairs by hydrogen bonds which thus maintain the form of the double helix. The structure of the bases is such that adenine must bond with thymine, and cytosine with guanine (Fig. 1.8). Subject to this limitation, the bases can occur in any sequence, and it is this freedom of arrangement which enables the chromosome to function as a bank of coded information.

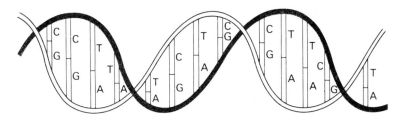

Fig. 1.8 Diagram of the Watson Crick model of the DNA molecule. A, adenine; C, cytosine; G, guanine; T, thymine.

The code is based upon triplet groups (codons) of the base units, giving 4^3 (64) different combinations, which is more than enough to code for the twenty common amino acids. The surplus is accounted for in part by several different codons coding for the same amino acid (referred to as degeneracy of the code), and in part by some codons acting as punctuation marks, coding for the beginning and end of a polypeptide.

Transmission of the information contained in the chromosomes requires an intermediary between these and the cytoplasm. This intermediary is ribonucleic acid (RNA), which, as we have seen, is similar in composition to DNA, except that uracil takes the place of thymine. RNA occurs in three main forms, one of which is messenger RNA (mRNA). This is synthesized in the nucleus, during the process called transcription, one of the two DNA strands acting as a template. The mRNA strand can then leave the nucleus carrying coded instructions which specify the assembly of amino acids into polypeptides and proteins, during the process called translation. Within the cytoplasm the mRNA strand attaches to a series of ribosomes, which are composed of proteins in association with a second type of RNA called ribosomal RNA (rRNA). The amino acids required are conveyed to the mRNA by a third type of RNA called transfer RNA (tRNA). This is composed of relatively small molecules, each of which can bond with one specific amino-acid molecule, and which carries also a base triplet (anticodon) which can bond only with the mRNA codon for that particular amino acid. For example, a tRNA molecule might carry the anticodon UUA (uracil,

uracil, adenine); this would enable it to bond with and thereby activate asparagine. The anticodon could only bond with the mRNA codon AAU (adenine, adenine, uracil), which is one of the codons for asparagine. This amino acid is therefore brought to precisely the point in the developing polypeptide molecule where it is required. The process can be visualized as a movement of ribosomes along the mRNA strand, with the developing polypeptide chain peeling off behind, and with more amino acids being conveyed to the growing end of the chain (Fig. 1.9). The polypeptides formed in this way contribute, either singly or in association, to form the enzymes or other proteins which characterize the activities of the cell concerned.

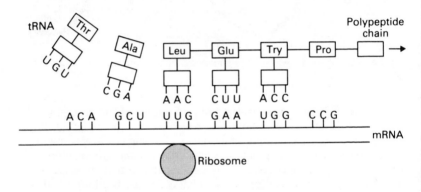

Fig. 1.9   Diagram of the synthesis of a polypeptide chain.

We are now in a position to attempt to define a gene. There are complications in this, but in principle a gene may be considered as being that part of the double helix strand which codes for a specific polypeptide. In other words, it is a sequence of base-pairs, which is sometimes termed a cistron. Such genes are distributed in linear fashion along the helix, together with other genes with different functions, including the synthesis of rRNA and tRNA, and the control of gene activities.

It will be recalled that chromosomes are paired; so also, therefore, are the genes. Each gene occupies a specific site (locus) on one chromosome of a pair, and has a corresponding gene at the corresponding locus on the other chromosome of the pair. The two members of the pair of genes may be identical (homozygous), so that they produce the same effect. But they may differ (heterozygous), producing alternative effects (for example, the normal pale form of the peppered moth, *Biston betularia*, or the black variety, *carbonaria*, which is favoured by natural selection in polluted areas in this country). Such different members of a gene pair are termed alleles (short for an earlier term, allelomorphs).

The existence of allelic pairs is a consequence of changes in genes called mutations, brought about, in their simplest form, by the substitution of one nucleotide base for another, with consequent change in the genetic code and in the product for which the gene is coding. Mutation is a recurrent phenomenon, and, because of this, and because of the accumulation of mutations in the genome (the complete set of genes), brought about largely by natural selection, allelic pairs are quite common, certainly much more so than was at one time thought. We shall see later that this confers a high degree of variability upon wild populations, providing a reserve upon which they can draw for adaptive responses to changes in their environment.

In some circumstances, especially when a number of characters have a close structural or functional relationship, a group of genes may act together as a super-gene. This condition is of importance in the control of polymorphism, which is defined as the existence in a population of two or more discontinuous forms of a species in such proportions that the rarest cannot be maintained by recurrent mutation (the implication of this being that the genes concerned must have been positively selected). The switch from one form (morph) to another may involve integrated characters determined by a number of genes, which can function most effectively by acting as a super-gene. Polymorphism in the snail *Cepaea* is an example of this; the colour of the shell, and the number, colour and fusion of colour bands, provide for adaptive responses to the colours of different types of background. Here at least four genes are jointly concerned. Another example is the existence of two types of flower (pin and thrum) in the primrose, *Primula vulgaris* (p. 175); here at least seven genes are concerned, cooperating in controlling style and pollen incompatibility, as well as morphological features.

The account so far given, while explaining the production of polypeptides, does not explain how this production is regulated in accordance with the functions required of particular cells. These requirements are in part determined by the positions and functions of the cells within the body, and in part by changes in the needs of the body from time to time; in accordance, for example, with external stimuli from the environment, or with different phases in the life cycle. A widely accepted model to explain this, derived initially from bacterial studies, but with wider applications, is based on the concept of the operon (Fig. 1.10). This is composed of one or more genes (structural genes) required for the synthesis of a particular protein, and an operator gene. Elsewhere, and not necessarily part of the operon, is a repressor gene, which is responsible for synthesizing

20

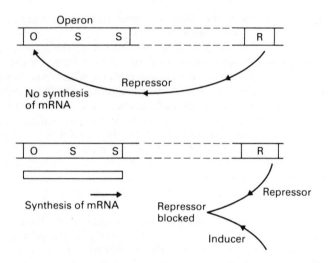

Fig. 1.10 Diagram of the regulation of gene action.

a repressor substance which binds to the operator gene and thereby switches off (represses) the mRNA-synthesizing activity of the operon. This repression can be evoked by the action of chemical signals arising either within the cell or from outside it. Alternatively, the signal substance may, by combining with the repressor, or in other and more complex ways, prevent it from binding to the operator gene, so.that synthesis of the appropriate mRNA is permitted (Fig. 1.10). Such a substance is called an inducer. Economy of energy is secured by this device, because enzymes required for the breakdown of a particular metabolite will only be produced when the metabolite is present and able to act, directly or indirectly, as an inducer.

An alternative possibility arises if the enzyme produced by an operon is yielding end-products in excess of what is required by the cell. These, it is supposed, can activate the repressor gene or its products, and thus bring about inhibition of the production of the enzyme. This is an example of the principle called feed-back, a term, derived from cybernetics, applied when the course of a reaction is influenced by the product of that reaction. In the example given, this is negative feed-back, for the product tends to reduce further output, but positive feed-back is possible in other contexts. This principle of feed-back is of fundamental importance and wide application in regulation, and in the maintenance of dynamic steady states, at all levels of biological organization, from the individual to the population (p. 65).

Cell Division and Differentiation

We have said enough to show that regulation of cell activities depends upon the continuous interchange of information between the nucleus and the cytoplasm. The nucleus is kept informed as to the state of the cell, while the cell is informed of the programme of action which it must follow, and is provided with the enzymes that it needs to do this. The same model helps us to secure some understanding of the complex events involved in the growth and differentiation of a multicellular organism by repeated cell division.

In bacteria there is usually a single ring-shaped chromosome which replicates as a whole, with simultaneous copying of each of the two strands of its double helix. A complete double helix can thus be passed to each daughter cell, with qualitatively equal division of the genetic material.

In eukaryote cells division of the genetic material takes place during a complex sequence of events termed mitosis, which can only be described here in brief outline (Fig. 1.11). This process, which does not occur in prokaryotes, is considered to begin when chromosomes become clearly visible within the nucleus by light microscopy (prophase). Prior to this, however, the two strands of each chromosomal double helix have become separated by the breaking of the hydrogen bonds uniting their base pairs. Two new strands are now synthesized, each complementary to one of the existing two, and thus two double helices are formed, each comprising one old strand and one new one. Thus each prophase chromosome is seen to be double, being composed of two chromatids united at a specialized region called the centromere, which itself later divides.

The nuclear membrane now breaks down and the double chromosomes, which have shortened and thickened, become oriented (metaphase) on the equatorial plate of a spindle which has been formed within the cytoplasm by the orderly arrangement of microtubules (p. 26). [In animal cells a body called the centriole (visible also within the non-dividing cell) is present at each pole of the spindle. Centrioles are not present in plant cells, but otherwise the whole process is essentially the same.] The sister chromatids of each chromosome now move apart (anaphase) towards opposite poles of the spindle (telophase), the movement resulting from some interaction between certain of the microtubules and the now divided centromeres. Each group of chromatids forms a new nucleus (telophase), nuclear membranes being generated in part from the remains of the parent nuclear membrane, but perhaps also from parts of the endoplasmic reticulum. The cytoplasm is now divided into the material of the two daughter cells, with an apportionment of

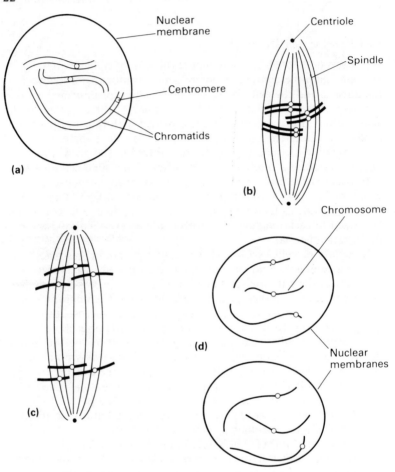

Fig. 1.11 Simplified diagram of the four stages of mitosis. (**a**) prophase, (**b**) metaphase, (**c**) anaphase and (**d**) telophase.

the organelles (p. 24), but not as precise a one as for the chromosomes.

The result of mitosis is thus a precise replication of genetic material, with distribution of identical information to each of the two daughter cells, carried in the same number of chromosomes as that of the parent cell. This precision is a necessary adaptation to ensure the continuance of the orderly functioning of the highly organized eukaryotic cell. But how, if this is so, is differentiation and division of labour brought about in the developing organism?

One fundamental factor, expressed in terms of the model already used, must be gene repression and derepression; the switching off and on of genes. Mitosis ensures that each cell in the organism contains the entire genome which the individual received from its parent or parents, but at any given time, in any particular cell, only part of this genome will be active. Some of its genes may be permanently switched off, in the sense that they will never function at any time during the life of the cell, or they may be only temporarily out of action. Perhaps in some instances the capacity of a gene to function is permanently lost, but this is certainly not always, perhaps not often, so, for there is convincing experimental evidence that genes which would not normally function in particular cells or tissues can be brought back into action if conditions suitable for this are provided.

For example, nuclei from the highly differentiated cells of the intestinal epithelium of larvae of the toad, *Xenopus laevis*, have been transplanted into enucleated eggs of the same species, and have then been found to promote normal development of the eggs into functional tadpoles and even sometimes into mature adults. Genes which would never have functioned within the environment of intestinal cells have thus been brought back into action by being placed in the quite different environment of a fertilized egg. It should not be inferred from this that genomes never undergo irreversible changes during differentiation; there is evidence that they may do, but that this is not necessarily always so.

To examine in any depth the processes that exert these influences upon the genome would take us too far afield. What we can say is that differentiation involves sequential gene action which translates coded genetic instructions into the enzymes that control the development, differentiation and metabolism of the cells, and that this sequential action cannot be determined solely by instructions built into the individual cell. This follows because the development of a cell, and the establishment of its efficient functioning, require that it be coordinated with neighbouring cells. Indeed, in plants, where the form and position of a cell is more rigidly fixed by its cell wall than in the freer environment of animal tissues, correct orientation can only be achieved through precise coordination with adjacent cells. These requirements can only be met by communication between cells, effected by the diffusion of nutrients or of the growth-regulating factors which we discuss later.

But this is only part of a wider spectrum of communication. Movement within the cell itself is clearly essential, effected by movement in both directions between nucleus and cytoplasm. By this means instructions are distributed from the nucleus, the activity of which is then modified by feed-back in the manner that we have outlined. Finally, and at the other extreme, there is information which is derived from stimuli arising in the environment, and which after suitable processing, is distributed by the nervous system in animals, and, as hormones, by diffusion in plants and through the blood system in animals. Thus the cells receive signals which evoke, throughout development and adult life, the responses that are most likely to ensure the survival of the organism and the transmission of information and instructions to its offspring. We refer later, in more detail, to these and other aspects of communication.

Cell Organelles

Within the cytoplasm of both plant and animal eukaryotic cells there are several structures, commonly membrane-bound, called organelles (Fig. 1.12). They comprise mitochondria, endoplasmic reticulum, Golgi bodies, microtubules, lysosomes, and (in plants) plastids. Mitochondria, Golgi bodies and plastids are lacking in prokaryotes, with a consequent dispersion of their functions. Electron microscopy shows mitochondria to be hollow rods or

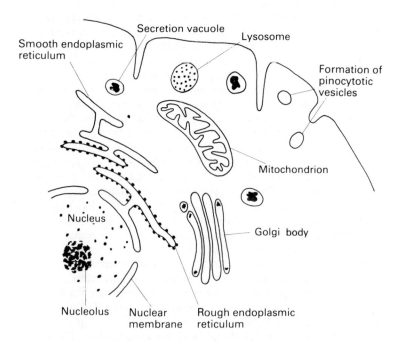

Fig. 1.12 Some features of the organization of an animal cell (not to scale).

spheres, bounded by two membranes, and with a highly characteristic structure. Both membranes contain protein and phospholipids. The outer one, which is permeable to six-carbon sugars and other large molecules, including ATP and ADP, permits metabolic exchanges between the mitochondria and the cytoplasm. The inner membrane is less permeable, and forms a boundary between the metabolic processes of the mitochondrion and those of the rest of the cell. It projects into the interior of the organelle to form tubular or disc-shaped structures called cristae, and bears small stalked bodies on its inner surface.

Mitochondria are often referred to as the powerhouses of the cell, and with good reason. Within them, and especially on and within the inner membrane, are located the enzyme systems of the tricarboxylic acid (TCA) cycle and respiratory chain (p. 38), which are of prime importance in energy conversion. At the ends of the stalked bodies are small spheres which contain ATP-ase; they are the sites for the synthesis of ATP, which passes into the cytoplasm and provides the energy for its activities. Probably mitochondria usually arise by division of pre-existing ones. They contain a protein-synthesizing system, including DNA and RNA, which is distinct from the main genetic system of the cell, and which would contribute to this power of self-reproduction. It has been estimated, however, that the mitochondrial system would not be sufficient to code for the whole of the structure of the organelle, which must therefore depend also upon the cooperative action of the nucleus and cytoplasm. Even so, it is far from clear how its complex structure is assembled. The evolutionary origin of mitochondria is no less obscure, but one challenging view, based upon their genetic capacity, is that they might have originated from independent bacteria-like organisms which invaded primordial cells and established permanent (symbiotic, p. 54) relationships within them. If this did indeed happen, it would have provided an early and dramatic demonstration of the important part played by interactions between different types of organism in the evolution of living systems (see also p. 27).

The Golgi bodies (or Golgi apparatus) are aggregations of hollow disc-like structures (cisternae), enclosed by double membranes, from which small vesicles may arise. The larger individual elements of the system are sometimes called dictyosomes, when they are separate from each other, but commonly the bodies are closely apposed to form stacks of membranes, easily recognizable in electron micrographs. The Golgi bodies are concerned in the process of secretion, which can be defined in this context as the production and release by a cell of materials (secretions) which have specific functions to fulfil. It is common, at some stage in the production of a secretion, for its constituents to be sequestered within Golgi cisternae, sometimes having first become visible within the rough endoplasmic reticulum (see below). It then leaves the cisternae within the smaller vesicles which are budded off from them, and it can then be transported to the upper (free) border of the cell, perhaps after further processing. Commonly the vesicle membranes become confluent with the plasma membrane, thereby permitting release of the secretion from the cell.

Closely associated with the Golgi bodies, and also enclosed by double membranes, is a system of canals which ramifies through the cytoplasm. This system, called the endoplasmic reticulum, connects with the nuclear membrane, and perhaps sometimes with the plasma membrane as well. This provides for a transport system, contributing to communication between the nucleus and the cytoplasm, and concerned also with secretion, as we have seen. Two forms of the endoplasmic reticulum are commonly present. In one, the smooth endoplasmic reticulum, the membranes have a smooth surface. In the other, the rough endoplasmic reticulum, the outer membrane carries on its outer surface many small granules (about 15 nm in size). These granules are composed of ribosomal RNA (rRNA), which we have seen to provide for one of the several stages in the synthesis of protein. Rough endoplasmic reticulum is thus especially conspicuous in cells that are manufacturing protein, being then visible as masses, although not in their finer detail, by light microscopy.

Hollow structures, called microtubules, formed of a polymerized protein (tubulin) appear in a variety of cells, and probably have diverse functions. They seem to be rigid, and to contribute to resist distortion of the shape of the cell. They perhaps provide a basis for various fibrous structures found in protozoans, and in certain rhizopods they probably have a major role in the positioning of the plates of the shell (test) during and after division. It has been thought that microtubules play some part in determining the orientation of the cellulose fibrils of plant cell walls, and they are certainly prominent in the thickened regions of the walls of the guard cells of stomata (p. 51).

Lysosomes are sac-like structures containing digestive enzymes. They feature prominently in intracellular digestion, both in the self-digestion (autolysis) of tissues, and in the digestion of food and other materials taken in by phagocytic cells.

Plant plastids

Characteristic of plant cells are a variety of bodies, often distinctively pigmented, called plastids, derived by division from

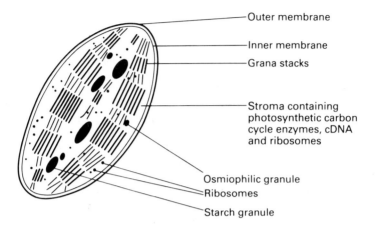

Outer membrane

Inner membrane

Grana stacks

Stroma containing photosynthetic carbon cycle enzymes, cDNA and ribosomes

Osmiophilic granule

Ribosomes

Starch granule

Fig. 1.13 Simplified diagram of a chloroplast.

smaller bodies called proplastids, which are transmitted through the ovum. The most important of these are the chloroplasts, which are green in colour because they contain chlorophyll (p. 35). These are the sites of photosynthesis. Those of higher plants (Fig. 1.13) are lens-shaped bodies up to 10 μm in length, of which a few or as many as 50 or more may be found in a single cell. They are bounded, like other organelles already mentioned, by a double membrane, and have a colourless granular matrix (stroma) containing a complex membrane system. This consists of double-membrane lamellae (thylakoids), stacked to form grana, which contain the chlorophyll and which are connected by intergrana. In lower plants the form of the chloroplasts may be more complex, but their structure is simpler; algae have chloroplasts with lamellae but no grana, while in the blue-green algae the lamellae lie free in the cytoplasm. No chloroplasts are present in photosynthetic bacteria (p. 37), their pigments being located on lipoprotein bodies called chromatophores.

The organization of chloroplasts recalls that of mitochondria, and there is, indeed, a close analogy. Both are essentially lipoprotein complexes, while chloroplasts, like mitochondria, possess their own DNA and RNA. Moreover, the inner membrane of the chloroplast contains not only the chlorophyll, but also its functionally associated enzymes, and it is the site of formation of ATP. The analogy has been pressed still further by the suggestion, for which there is thought to be good evidence, that chloroplasts evolved from an invasion of primordial cells by blue-green algae. However this may be, the possession by chloroplasts of a genetic system suggests, again as with mitochondria, a capacity for self-replication, but origin *de novo* cannot be excluded.

The development of chloroplasts from proplastids involves the latter budding off vesicles from which the lamellae are derived, chlorophyll being laid down in the thylakoids. In flowering plants this development is only completed in light, and is arrested in darkness. In gymnosperms (p. 167), however, light is not an essential requirement. Development is also affected by the properties of particular cells; an illustration of the adaptive cell differentiation which we have already mentioned as a feature of plant and animal organization. For example, chloroplasts occur chiefly in parts of the plant that are exposed to light, which are the leaves and also the young stems. They are characteristic also of the internal tissues (parenchyma and collenchyma) of the leaves, their development being arrested, in some way not understood, in the epidermal cells of the leaf surface.

Other types of plastids also occur in plant cells. Chromoplasts are coloured bodies, as their name implies; they contain pigments, including carotene and xanthophylls, which give them a red, orange or yellow colour. Other plastids, called leucoplasts, are colourless. The function of the chromoplasts is not clear, but leucoplasts synthesize and store food reserves, particularly starch, fats and oils, and perhaps protein as well. All of these bodies are derived from maternally transmitted proplastids. It is possible that they are formed along the same developmental path as are the chloroplasts, with their development being arrested at an appropriate stage. One example of the close interrelationship between the several types is the development of chlorophyll in the starch-storing amyloplasts of potato tubers when they are exposed to light. Another is the change in colour during the ripening of the fruits of peppers and tomatoes, which is due to the replacement of the chlorophyll of the chloroplasts by red xanthophylls.

Streaming and Cyclosis

We have seen that intracellular membranes form a flexible and integrated transport system, which is all the more flexible because of the possibility that one type of membrane can be transformed into another. This, however, is not the only transport mechanism available to the cell, for materials can also be moved by cytoplasmic streaming. Two examples illustrate this.

Amoeba, which, although a complete protozoan organism, can be regarded for this purpose as equivalent to a cell, shows continuous streaming within its cytoplasm, and, indeed, this is part of its locomotor mechanism. Many plant cells also show a characteristic rotation of the cytoplasm, called cyclosis, which can be seen to carry cell materials with it.

The rotatory movement of cyclosis is associated with the presence of one or more large vacuoles in the cytoplasm (protoplast) of plant cells, each vacuole containing water with salts and inorganic materials in solution (cell sap), and being bounded by a membrane, the tonoplast. This membrane, like the plasma membrane, has a differential permeability, permitting the passage through it of water and a limited range of solutes. If the cell is surrounded by a fluid of lower osmotic pressure than its contents, the cell is said to have a higher osmotic potential (p. 111); water passes into it by osmosis and builds up a pressure in the vacuole called turgor pressure. A limit to this is reached when the pressure exerted by the relatively inelastic cell wall counterbalances the pressure built up against it by the cell membrane. Molecules of water will still continue to pass into and out of the vacuole, but a state of equilibrium has been reached at which net entry of water has been stopped.

The equilibrium will be affected by any circumstance which brings about a fall in turgor pressure. This happens, for example, when loss of water from a plant by evaporation exceeds the uptake of water by the roots, producing a state of water deficit. The protoplast remains pressed to the cell wall, but the reduction in pressure causes the walls to collapse inwards, and the plant is said to be wilted. Reduction in turgor pressure also results if cells are placed in a hypertonic solution (i.e. one which has a higher osmotic pressure than that of the vacuolar fluid). The water in the cell is now at a higher concentration than the water outside it, and may be considered to have a greater free energy. In consequence, water will pass out from the vacuole more quickly than it can enter, and a net osmotic loss of water ensues. This results in plasmolysis (Fig. 1.14), in which the vacuole shrinks and the protoplast draws away from the cell wall. Recovery occurs in both instances if conditions are reversed, provided that the loss of water has not proceeded too far.

It will now be evident that interaction between the plant cell wall

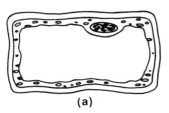 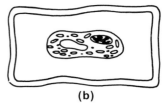

(a) (b)

Fig. 1.14 Diagrammatic comparison of a normal plant cell (**a**) with a plasmolyzed one (**b**).

and the central vacuole not only maintains the turgor pressure of the individual cell, but also makes an important contribution to the support of the plant tissues, while the critical importance of water is also apparent. This interaction also facilitates the survival of the plant cell in a greater range of osmotic conditions than is possible for animal cells, which are much more exposed in this respect. It is because of this that animals have had to deploy diverse and often complex osmoregulatory systems.

Ergastic Substances

We have mentioned the presence of food reserves in plastids, but in all types of cell, both in plants and in animals, metabolic products may be found lying free in the cytoplasm. Two examples of this are starch in plant cells and glycogen in animal cells. Fatty, proteinaceous and mineral materials are also common, these being amorphous, crystalline or liquid, according to the nature of the deposit. Waste products of metabolism may also be present, more particularly in senile cells; they need careful assessment by the observer, for the boundary between waste materials and adaptively useful ones is not always easy to draw. In general, all metabolic products laid down either temporarily or permanently within the cell are termed ergastic substances.

2 Life and Energy

The Origin of Life

The complex organization of living material makes it difficult to understand how it could ever have become established. Was its origin governed by the same evolutionary processes as have determined the progressive diversification of living organisms, or was it governed by other factors that take the question beyond the reach of scientific analysis? These are matters that can here be dealt with only very briefly, but it is essential to make some reference to them, not only because of their inherent interest, but because they help us to understand better the different patterns of organization of living systems, and the relationships by which these systems are linked with each other and with the physical environment. It will be prudent, however, to remember that there is no *a priori* reason for supposing that the human mind has the capacity to penetrate into the full depth of the problem.

Our sun and its planetary system are thought to have arisen, perhaps 4600 million years ago, by condensation within a cloud of gas consisting largely of hydrogen, which, with a small percentage of helium, is the predominant element of the universe, and still makes up some 80% of the sun's matter. It is likely that this gas would also have contained particles, especially of silicates, to which could have been adsorbed hydrocarbons and other simple carbon compounds. These compounds, and the elements of which they are composed, are widespread in the universe, and it has been remarked as a peculiar circumstance that amongst the most abundant of them are those that have proved to be the best suited for the establishment of life upon the earth. (Oxygen, however, is an exception, for a reason which we shall see.)

The formation of these elements from primeval hydrogen could have taken place through nuclear reactions in the stars, and especially in the explosions that are responsible for supernovae. The energy of electrical discharges and of ultra-violet light would have promoted their chemical combination, reactivity being increased by their adsorption to particulate surfaces. These reactions, which have been found to occur under comparable conditions in the laboratory, would have continued upon the earth, aided by its very high temperature at that time. This could have led (and again there is supporting evidence from the laboratory) to the formation of amino acids and the combination and polymerization of these and other small molecules to form the larger molecules of the carbon compounds upon which the development of life was to depend. We customarily describe these carbon molecules as organic, because they were initially identified in living organisms, and are still characteristic products of them, but it will be apparent that on this analysis their first appearance must have preceded the appearance of life.

These propositions cannot be discussed in detail here, and it is all the more important not to underestimate the complexity of the questions that they raise. One aspect, however, that we can readily see is that life today is particularly well-suited by marine conditions (pp. 112, 128). It is therefore supposed that primaeval oceans, formed as materials were swept from the land into the water that was condensing on the cooling surface of the earth, provided a reactive medium in which complex molecules became aggregated together into self-replicating systems. It is further supposed that these acquired identities distinct from the surrounding environment by developing surface membranes, and that they began to maintain themselves by exchanges with that environment that were the beginnings of metabolism. And so life began on the earth, probably within 1000 million years of the formation of the solar system, for microfossils resembling bacteria are found in sedimentary rocks 3200 to 3400 million years old. The first eukaryotes appear much later, in deposits about 1500 million years old. It has been surmised that their evolution may have had to await the establishment of mitochondria (p. 25).

Autotrophy and Heterotrophy

The proto-organisms that we are envisaging would have needed a supply of energy to drive them. It is supposed that they would have obtained this by taking up complex molecules already produced in the environment by solar radiation and electric discharges, and breaking these down by chemical reactions to release their potential energy. What can we infer as to the nature of these reactions?

Living organisms today derive their energy in one of two ways. One of these is autotrophy, which is the ability to manufacture all of the energy-rich organic compounds that they need from inorganic sources (water, carbon dioxide and simple mineral salts). Autotrophs may be phototrophs, capturing the required energy from sunlight, or chemotrophs, using energy obtained by oxidation of inorganic compounds without recourse to sunlight. The other method of securing energy is heterotrophy, which differs fundamentally from autotrophy in that the energy is obtained from complex carbon compounds taken into the body from outside. In both cases the potential energy of the macromolecules is released by catabolic reactions which are usually dependent on a supply of oxygen (aerobic metabolism).

The earliest organisms that we have been postulating were, in this terminology, primitive heterotrophs, for we are arguing that these, too, were obtaining their energy by the breakdown of complex molecules which were already present in their environment. But there was a fundamental difference from present-day conditions, because their environment must have been a reducing one, rich in free and combined hydrogen. Life was evolving in the absence of free oxygen, by means of energy released through anaerobic metabolism. By what metabolic pathways was this achieved?

Glycolysis

To find the answer to this question it must first be understood that, although the life that surrounds us today depends ultimately on the presence of oxygen, it is nevertheless possible for it to be maintained, to varying degrees and often in exceptional conditions, by anaerobic metabolism. This is because the release of energy by oxidative reactions does not necessarily require oxygen to be present. Oxidation, defined chemically, is the loss of electrons, which is often associated with the transfer of hydrogen. Thus biological metabolites can be oxidized, with the release of energy, by acting as hydrogen donors, in reactions mediated by enzymes called dehydrogenases. The hydrogen is transferred to substances called hydrogen acceptors or carriers, which are said to be reduced thereby.

Anaerobic oxidation is effected in present-day organisms by a universally distributed chain of reactions that constitute the process called glycolysis. Its universality indicates that it must have been established very early in evolution, for it is unlikely that so complex a sequence of reactions would have arisen more than once. It is therefore reasonable to assume that primitive heterotrophs also made use of glycolysis. In simplified outline, the glycolytic (anaerobic) metabolism of glucose proceeds today as follows (Fig. 2.1).

The glucose molecule must first be phosphorylated to glucose-6-phosphate; this is then converted to fructose-6-phosphate, which is further phosphorylated to fructose-1:6-diphosphate. These phosphorylations, which are essentially a priming process, depend upon the conversion of two molecules of ATP to two molecules of ADP. At this stage, then, glycolysis is actually using up energy stores. In a further sequence of reactions, each 6-carbon carbohydrate molecule is split into two 3-carbon molecules of triose phosphate,

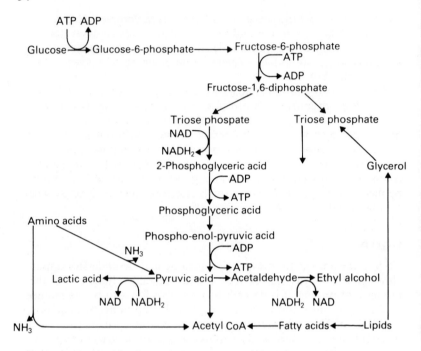

Fig. 2.1 Outline of the sequences of reactions involved in glycolysis and fermentation.

each of which is broken down by stages to a molecule of pyruvic acid. During these later stages four molecules of ATP are generated; sufficient to compensate for the two used in the preliminary phosphorylations, and to provide a net gain of two molecules of ATP from the breakdown of one glucose molecule.

$$C_6H_{12}O_6 \rightarrow 2CH_3CO.COOH$$
$$2NAD \rightarrow 2NADH_2$$
$$2ATP \rightarrow 2ADP$$
$$4ADP \rightarrow 4ATP$$

Thus there is a total energy gain of 2×30.56 kJ $= 61.1$ kJ (p. 8).

It will be noted that four atoms of hydrogen are also released. These are taken up by two molecules of a hydrogen carrier, the coenzyme (p. 9) nicotinamide adenine dinucleotide (NAD^+), which are thereby reduced to $NADH_2$. Under anaerobic conditions these hydrogen atoms can be used to reduce pyruvic acid to lactic acid, notably in muscle and other animal tissues. The energy yield of this reaction is very small (25.1 kJ per molecule of pyruvic acid), but it is sufficient to permit, for example, the continued contraction of vertebrate muscle when oxygen is being used up faster than it can be replaced. However, this anaerobic phase of contraction, during which the animal accumulates what is called an oxygen debt, is of brief duration, for fatigue develops as a result of the accumulation of the lactic acid:

$$CH_3CO.COOH + NADH_2 \rightleftharpoons CH_3CHOH.COOH + NAD$$
$$\text{(pyruvic acid)} \qquad\qquad \text{(lactic acid)}$$

Many bacteria and fungi, including yeast, carry out a form of anaerobic catabolism called fermentation, agreeably exploited by man, in which the pyruvic acid is reduced to ethanol, with the production of CO_2, and with acetaldehyde as an intermediary:

$$CH_3CHO + NADH_2 \rightarrow CH_3CH_2OH + NAD$$

$$CH_3CO.COOH \qquad \rightarrow CH_3CHO + CO_2$$

Photosynthesis

Glycolysis, which has a relatively poor yield of energy, is, for most organisms, only part of their system of oxidative reactions; nor could it ever have been adequate for the permanent establishment of life, for the primitive heterotrophs would have used up their nutrients at a greater rate than they could have been replaced, and would eventually have exhausted their supply. Life would have come to an early end had it not been for a fundamentally important advance: the establishment of autotrophy in the form called photosynthesis, which is characteristic today of the green plants and of certain bacteria. Photosynthesis was made possible by the evolution of the green pigment chlorophyll, with its ability to capture the energy of the sun. It probably appeared first in anaerobic bacteria (precursors, presumably, of the photosynthetic bacteria to be mentioned later) and must have been strongly promoted by natural selection because of the improved prospect of survival which it afforded to the organisms able to practice it. Aerobic photosynthesis would therefore soon have followed, in organisms which would have been precursors of the blue-green algae.

Chlorophyll is a coordination complex of magnesium and the cyclic tetrapyrrole derivative called porphyrin, which consists of four pyrrole rings joined by methane bridges (Fig. 2.2). Porphyrin is a molecule of remarkable potentialities, which is of virtually universal distribution in living organisms, and which gives rise to diverse compounds formed along similar biosynthetic pathways. It can combine with iron instead of with magnesium, and in this form it occurs in the respiratory pigments (cytochromes) of the electron

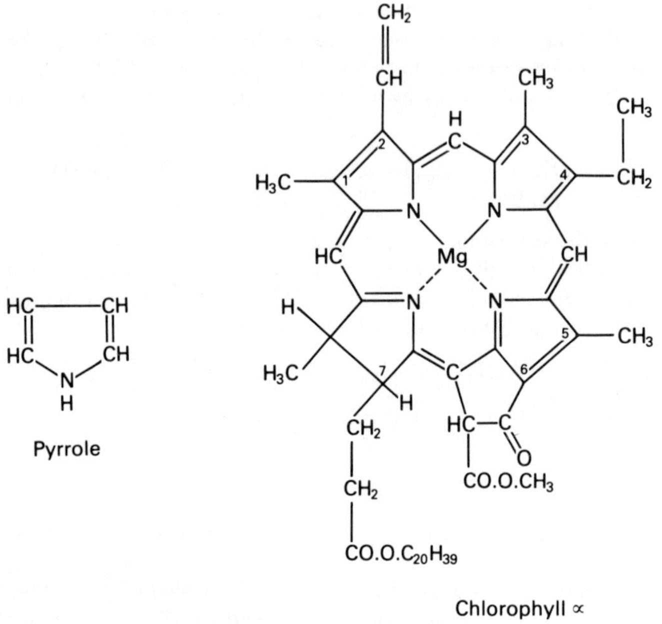

Fig. 2.2 Pyrrole ring and chlorophyll α. Chlorophyll β has CHO substituted at position 3.

transport system, to be described shortly, and also provides the prosthetic groups of various enzymes, including oxidases and peroxidases. An iron-porphyrin complex is also the basis of the oxygen-carrying pigment, haemoglobin, found in vertebrates and a number of invertebrates as an essential part of their respiratory systems (p. 101).

There are a number of types of chlorophyll, the most abundant being chlorophyll *a* and *b*, both of which occur in the higher plants and the green algae. Their main absorption peaks are at about 429 and 453 nm respectively, in the blue-violet region, with subsidiary peaks at about 660 and 642 nm respectively, in the red region. They are associated in the chloroplasts with two carotenoid pigments, mainly orange-red β-carotene and various xanthophylls. These lipid compounds have absorption spectra which lie between the two bands of chlorophyll. They do not themselves carry out photosynthesis, but they protect the chlorophylls from photo-oxidation, and it is supposed that they increase the efficiency of their photosynthetic action by absorbing light energy and transferring it to them.

In the green plant

The amount of energy absorbed by green plants is less than 1% of the solar energy reaching the earth, for about 60% of this is immediately reflected, most of the remainder being absorbed by water, land and air, and re-radiated as low energy heat. Given an estimated annual solar energy input of 2.1×10^{24} J $(= 0.5 \times 10^{24}$ cal), the conversion efficiency is of the order of 0.16%. The radiant energy actually used in photosynthesis reaches the green plant within the wavelength range of about 400–700 nm; this corresponds closely with the range (380–760 nm) which we normally perceive as light, and which stimulates the photoreceptors of animals in general. It is received as discrete units or quanta which are absorbed by the molecules of the chlorophyll pigment and thereby activate them. This activation energy is used, with water as hydrogen donor, to reduce carbon dioxide and thereby build up carbohydrates and other organic molecules. The overall equation for carbohydrate formation may be written thus:

$$CO_2 + 2H_2O \rightarrow (CH_2O) + H_2O + O_2$$

The gain in free energy is 490 kJ (117 kcal). Little free glucose is produced, the main carbohydrate products being sucrose and starch.

The reactions involved in the process, localized entirely within the chloroplasts, take place in two main stages, one of these (the light phase) being photochemical (light-dependent) while the other (the dark phase) is enzymatic (light-independent). During the light phase, which takes place in the grana of the thylakoids, water molecules are split into H^+ and OH^- ions (photolysis). Because of this, the water molecules evolved by the reaction are different from those that entered it, the latter being the source of the evolved oxygen. The photolysis results in a flow of electrons along two distinct reaction pathways, light energy being converted to chemical energy by the formation of ATP and reduced NADP ($NADPH_2$; reduced nicotinamide adenine dinucleotide phosphate). The assimilation of CO_2 occurs in the dark phase, which takes place in the stroma of the chloroplasts. The stored energy of ATP is used to drive the reduction of CO_2 to carbohydrate, the hydrogen needed for the reduction being provided by the reduced NADP.

In bacteria

Certain autotrophic bacteria carry out photosynthesis by means of modified forms of chlorophyll called bacteriochlorophylls, which differ from the pigments of green plants in absorbing light of a longer wavelength and with less energy. Further differences are that these bacteria do not use water as hydrogen donor, they do not evolve oxygen during photosynthesis, and they often have chemical requirements additional to water and carbon dioxide. Three main groups are recognized:

1. The green sulphur bacteria (Chlorobacteriaceae) are strictly autotrophic anaerobes which contain carotene in addition to bacteriochlorophyll. They use hydrogen sulphide as hydrogen donor, and oxidize it typically to sulphur (but sometimes to sulphate).

$$CO_2 + 2H_2S \rightarrow (CH_2O) + H_2O + 2S$$

It is thought that these organisms, which are common where H_2S is being produced by organic decomposition, may have been an important source of natural elemental sulphur.

2. The purple sulphur bacteria (Thiorhodaceae) contain a brownish-red carotenoid in addition to bacteriochlorophyll. Most are strictly autotrophic anaerobes, using hydrogen sulphide as hydrogen donor, and oxidizing it to sulphur and sulphate. Some, however, can use molecular hydrogen, while some may use simple organic (fatty) acids as a source of hydrogen. They commonly occur in warm springs in volcanic regions:

3. The purple non-sulphur bacteria (Athiorhodaceae) have similar pigments to the purple sulphur group, but their nutrition is more complex. They grow anaerobically in light, and can then fix carbon dioxide and often nitrogen, some utilizing molecular hydrogen or inorganic sulphur compounds. Preferentially, however, they use simple organic compounds as hydrogen donors, and they also need certain others for their growth, so that although they are photosynthetic they are not strictly autotrophic. Some members of the group can grow aerobically in the dark, and are then, of course, heterotrophic.

Bacterial photosynthesis is very small in amount in comparison with the activities of green plants, but it is of great evolutionary interest. It is, however, uncertain whether it represents stages in the evolution of the complex and highly organized photosynthesis of higher plants, or whether it has arisen independently.

The Tricarboxylic Acid (TCA) Cycle

The full establishment of photosynthesis led to the accumulation of oxygen in the atmosphere in amounts which significantly influenced further evolution in two ways. Prior to its appearance, the surface of the earth must have been intensely irradiated with the ultra-violet component of the solar spectrum, which has a damaging effect on nucleic acids and proteins. Presumably, then, the earliest forms of life must have sheltered from this in some way, perhaps well below the surface of the waters. Once oxygen began to accumulate, however, the ultra-violet radiation produced an ozone layer from it high in the upper atmosphere. This layer would have shielded the earth, as it does today, from the full intensity of the damaging radiation, and would thus have allowed primordial organisms some freedom to move and expose themselves. Concern that chlorofluorocarbons may erode this layer is beginning to lead to international control of their use as aerosol propellants.

This accumulation of oxygen also made possible the further evolution of metabolic pathways. It was now possible for the breakdown of pyruvic acid to be extended by a sequence of reactions which gave a greatly increased yield of energy. These reactions, in contrast to those of glycolysis, are oxygen-dependent (aerobic), and it is worth remembering that their continued operation depends absolutely upon the release of oxygen into the atmosphere during the photosynthetic activity of green plants. This relationship is thus fundamental to the maintenance of the environment to which life, as we know it, is adapted.

The sequence of reactions (Fig. 2.3; fuller accounts will be found

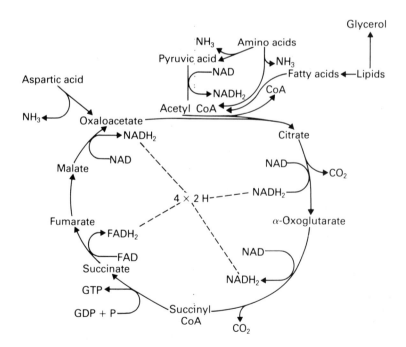

Fig. 2.3 Outline of the sequence of reactions involved in the tricarboxylic acid (TCA) cycle.

in biochemical texts) begins with the oxidation of pyruvate, the acetyl radical of which condenses with an important sulphur-containing agent, co-enzyme A, to form an activated (high-energy) compound, acetyl coenzyme A (acetyl-CoA). Carbon dioxide is liberated, together with hydrogen atoms (two from each molecule of pyruvate), which are taken up by NAD to form $NADH_2$. Acetyl-CoA now enters a self-regenerating sequence of enzyme-mediated reactions which constitute the tricarboxylic acid (TCA) cycle, often called the Krebs cycle (after the biochemist who was mainly responsible for its elucidation) or the citric acid cycle. Acetyl-CoA condenses with 4-carbon oxaloacetate to form 6-carbon citrate, the acetyl group being regenerated. There follows a series of four oxidations (dehydrogenations) from which eight hydrogen atoms and two molecules of CO_2 are produced. The cycle, during which three molecules of water are used, is completed by the regeneration of oxaloacetate. Three of the dehydrogenations require NAD as coenzyme and yield $NADH_2$, but the remaining one, in which succinate is oxidized to fumarate by succinic dehydrogenase, uses flavin adenine nucleotide (FAD), the prosthetic group of succinic dehydrogenase, and yields $FADH_2$.

The Electron Transport System

The sequence of chemical reactions by which free energy is released in the cell constitutes respiration (or cell respiration, see p. 93). Up to this point the reactions are anaerobic. There now follows an aerobic phase during which the reduced NAD and FAD are re-oxidized. The hydrogen atoms which they have taken up (Fig. 2.3) are split into hydrogen ions and electrons. The former are released into the cytoplasm, while the electrons are passed into a chain of electron carriers which can readily take up or lose an electron. These carriers, which include the cytochromes mentioned earlier (p. 10), constitute the electron transport system or respiratory chain.

The tendency for any substance to gain or lose an electron is expressed as its oxidation-reduction (redox) potential, those with a high redox potential being able to oxidize those with a lower potential. Within the respiratory chain, illustrated in simplified outline in Fig. 2.4, the electrons are enabled to flow down a gradient of redox potential. As they do so, the successive changes of potential lower the energy level of the electrons, the energy thus released being used at three points in the chain to generate ATP from phosphate and ADP. This process is called respiratory chain phosphorylation, in contrast to the substrate-linked phosphorylation of glycolysis. The hydrogen ions finally combine with oxygen, so that water is formed

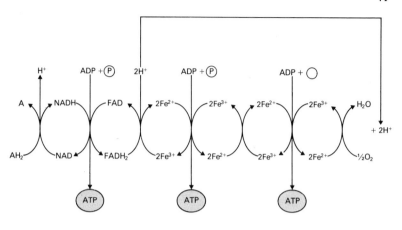

Fig. 2.4 Outline of the sequence of reactions involved in the functioning of the respiratory chain. A, substrate.

at the end of the chain, the overall equation for the complete oxidation of pyruvic acid being:

$$2CH_2CO.COOH + 5O_2 \rightarrow 6CO_2 + 4H_2O + \text{free energy}$$

Respiratory Energetics

We can now combine the results of glycolysis and of the action of the TCA cycle and respiratory chain to calculate the total energy yield for the complete oxidation of one molecule of glucose. It will be recalled that the breakdown of one molecule of glucose to two molecules of pyruvate yields two molecules of ATP by substrate-linked phosphorylation. It also yields two molecules of $NADH_2$. These, given aerobic conditions, are oxidized back to NAD, yielding six molecules of ATP. The subsequent oxidation of one molecule of pyruvate yields 15 molecules of ATP, derived as shown in Table 2.1, the formation of the ATP being mediated by the respiratory chain enzymes except for the formation of succinate from succinyl-CoA. Thus one molecule of glucose generates 38 molecules of ATP. Since each molecule of ATP yields 30.56 kJ, the total energy yield is 38×30.56 kJ = 1161 kJ (= 277.4 kcal). It follows that the organism is functioning with a conversion efficiency of about 40%, for the theoretical yield from one molecule of glucose is about 2872 kJ (684 kcal).

Looking at these aspects of energetics in more general and evolutionary terms, the whole process shows first of all the great

Table 2.1 Yield of ATP from the complete oxidation of one molecule of glucose via the TCA cycle. To the values shown must be added the net yield of two molecules from glycolysis, giving a grand total of 38. (Oxidation of one molecule of $NADH_2$ by O_2 yields three molecules of ATP, but $FADH_2$ yields only two.)

Reaction	Number of ATP molecules formed
Glucose to pyruvate	
(Triose phosphate to 2-phosphoglyceric acid)	3
Pyruvate to acetyl CoA	3
Citrate to α-oxoglutarate	3
α-oxoglutarate to succinyl CoA	3
Succinyl CoA to succinate*	1
Succinate to fumarate	2
Malate to oxaloacetate	3

* Guanosine triphosphate (GTP) is equivalent to ATP in energy value.

advantage which organisms obtain by using ATP as their energy currency. The most obvious is that the energy, gained in small amounts in a sequence of stages, can also be released in small amounts. This avoids any destructive action resulting from energy release, while a fine control can be applied to the energy output through the regulating action of enzymes.

But other important consequences flowed from the establishment of oxygen-dependent phosphorylation. Obviously it increases, in comparison with glycolysis, the energy yield from the breakdown of glucose, but this is not all, for the oxidation of pyruvate is only one of the ways in which acetyl-CoA can be generated. It is also formed by the oxidation of fats, and of many amino-acids (Fig. 2.3). Because of this, the TCA cycle and respiratory chain form a common pathway for the oxidation of many organic substances, and can also provide for the interconversion of some of these materials, while contributing also to their synthesis.

Secondary Heterotrophy

Evidently, then, the establishment of these pathways resulted in greatly increased metabolic efficiency, and this must have promoted the rapid increase and diversification of living organisms, accompanied by competition between them for the limited resources of their environment. This, as much as anything, may well have been the crucial factor leading to the evolution of a new type of organism which began to exploit the autotrophs by ingesting them and thus securing the energy store of the organic molecules that they had manufactured, instead of having to make its own. These new organisms were thus heterotrophs. In the light of our earlier argument, and to distinguish them from the primitive anaerobic heterotrophs, we must call these organisms secondary heterotrophs. They are represented today by the animal kingdom; normally aerobic organisms, that interlock with the life of plants, and are totally dependent upon them, in ways that are fundamental to the design of the life of our planet.

Animals, however, are not the only surviving heterotrophs. So also are many bacteria, and also the fungi, both of which groups play a fundamentally important part in the economy of nature. This is because animals and plants, by death and by the discharge of metabolic products, create a massive problem that is not so different in principle from human pollution of the environment, although man has proved more fertile in inventing ways of achieving this. Packaging, for example, is said to account for 30–40% of municipal solid waste in the United States. But the very term 'waste' is a peculiarly human concept, for the living world depends upon continuous cycling of materials in which nothing is wasted.

Pollution can be defined in broad terms as the consumption of environmental quality (see Cottrell, *Environmental Economics*, in this series); it contributes, therefore, to the irreversible fall in the quality of energy that is expressed in the Second Law of Thermodynamics. The energy can be replaced from solar radiation, but essential materials such as nitrogen, carbon and phosphorus cannot. They must be recycled, and, if this were not achieved, life would come to an end, choked by its own products. This recycling is made possible by the activities of a variety of organisms which maintain the fertility of the soil on land (and similar principles obtain in aquatic habitats, see Fig. 4.1, p. 66). It is exemplified by the carbon cycle, illustrated in simplified form in Fig. 2.5. We ourselves make a significant contribution to this by our combustion of fossil fuels (pp. 53, 72), which, it has been claimed, may have increased the level of CO_2 in the atmosphere by as much as 15% in the last hundred years. Here, however, as with other nutrient cycles (Fig. 3.5), it is the contribution of bacteria and fungi which is of the utmost importance, as we see in more detail in the next chapter.

Bacteria, averaging about 1 μm in diameter, and comprising over a thousand species, show an immense physiological and biochemical diversity. This matches their ubiquitous distribution, which is aided by their resistance, partly through spore formation, to conditions which would destroy most other organisms. We have already seen that some are photosynthetic. Others, presumably derived, like blue-green algae, from photosynthetic ancestors, are heterotrophic, breaking down organic materials to ammonia, CO_2, sulphates and phosphates. In this way they function as decomposers, contributing

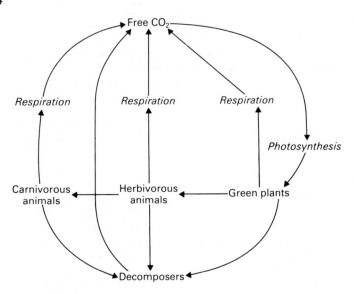

Fig. 2.5 Outline of the carbon cycle.

to the decay of organic residues, helping to maintain the fertility of the soil by their metabolism of nitrogen, and releasing nutrients into the sea at all levels, from the surface water to bottom deposits. Thus they contribute to the completion of the cycling of nutrients upon which the life of this planet depends (p. 59). While, therefore, they force themselves on our attention as agents of disease, they are, as a group, primarily beneficial. Furthermore, some of their activities have lent themselves to exploitation on an industrial scale, in the manufacture, for example, of products as diverse as vinegar, lactic acid, vitamin B_{12} and streptomycin. Their use as decomposers in sewage beds is another aspect of this exploitation.

To this must be added the immense potentialities of recombinant DNA techniques, an aspect of genetic engineering. These depend upon the use of enzymes for cleaving and rejoining DNA sequences. Foreign DNA can thus be combined with the DNA of a bacterial cell (by insertion into the plasmids of *Escherichia coli*, for example; a bacterium which lives harmlessly in the human intestine and is widely used in laboratory work). This gives the cell new genetic capacities which can be brought into rapid production (*E. coli* divides once every 15–20 min). Amongst the possibilities of genetic engineering that are now envisaged, although these still await full exploration, are the bacterial production of vertebrate polypeptide hormones such as insulin (p. 215) and the improvement of the effectiveness of nitrogen fixation by symbiotic systems (p. 55).

Fungi (Mycota), a group including rusts, mildews, yeasts and toadstools, and containing at least 50 000 species, with doubtless many more still to be identified, are closely associated with bacteria in some of these respects. Although they are often considered as plants, a case can be made out for regarding them as a separate kingdom, distinct from both plants and animals, partly because they are heterotrophs, and partly because most of them contain chitin, which is primarily an animal characteristic. Their origin, too, is obscure, but it was probably from flagellate algae. Perhaps, too, they are polyphyletic, having evolved along more than one line of ancestry.

Some fungi are unicellular (yeasts are an example). Others are multicellular (Fig. 8.2), with bodies formed of threads (hyphae) grouped into a mycelium from which spores develop, either directly or in fruiting bodies, as in mushrooms. Some fungi are saprophytes (feeding on dead organic matter), breaking down their food by means of enzymes that they secrete onto it, the products being absorbed through their hyphae, or sometimes through specialized regions of these. Their carbon is obtained from sugars or starch, and their nitrogen from the decomposition of proteins, or from ammonia and nitrates. It is thus that they contribute, with bacteria, to organic decay. Other fungi are parasites. This term is applied to organisms which live on or in others (their hosts) and feed at their expense, remaining closely associated with them and depending upon them for essential nutrients. In doing so, they may or may not harm them. The hosts of parasitic fungi are sometimes animals but usually plants, which makes them, when harmful, immensely important in human cultivation. Some fungi have carried their heterotrophy to the point of evolving remarkable adaptations for trapping animals. One such device involves the development from the hyphae of loops or rings which permit a round worm (nematode) to enter. When stimulated by the prey, these loops swell and trap it, branches of the fungus then entering the worm and releasing digestive enzymes.

Fungi, like bacteria, have proved of great importance for our technology, notably in the production of antibacterial substances (antibiotics) which weaken the bacterial cell surface and make the organism vulnerable to adverse environmental influences. Penicillin is a familiar example.

3 Primary Production

Lower Plants

The fundamental differences in structure and organization which distinguish animals from green plants are determined primarily by modes of nutrition. Animals are typically mobile, for they have commonly to seek their food and, even when they are fixed in position, must actively collect it. Only when they live within other organisms, and not always then, can they sink into a state of passive dependence upon their partner, and reduce to some extent the locomotor and behavioural adaptations which are so characteristic a feature of animal life. The nutrition of green plants, with its demand for radiant energy, carbon dioxide, water and certain solutes, presents a different set of problems. Their history, initiated as a floating mode of life in the sea, has tended towards a sessile and rooted life on land, adapted for the uptake and transport of nutrients, without diffusion of energy in exploration and pursuit.

The requirements for plant life are seen at their simplest in the aquatic algae. They range from minute unicellular forms to multicellular ones of substantial size, but, with nutrients in the water that surrounds them, and supported by its buoyancy, even the multicellular algae (Fig. 8.4) lack the vascular and supporting tissues that are necessary for the higher groups of terrestrial plants. There are, of course, structural specializations in the larger algae, some of which are very familiar on the sea shore. *Fucus vesiculosus* (Fig. 3.1a), conspicuous on the middle shore, is aided in its support by a thickened midrib and air bladders, while *Laminaria* (Fig. 3.1b), exposed at low tide, has its thallus differentiated into a holdfast for attachment, a stalk (stipe), and a blade which in related Pacific genera may reach a length of nearly 60 m. All of this, however, is achieved at only a modest level of organization. Lacking differentiation into root, stem and leaf, their whole body is constituted by a relatively simple and undifferentiated thallus, the movement of nutrients, in the absence of specialized conduction pathways, being effected by diffusion from cell to cell. Nevertheless, the ecological impact of the group is impressive. It is estimated that about 33 % of the energy fixation carried out on earth is effected in the oceans, mainly by algae, and, in particular, by the unicellular

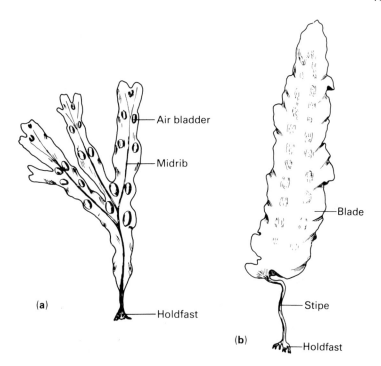

Fig. 3.1 (**a**) *Fucus vesiculosus*, bladder rack (× $\frac{1}{8}$). (**b**) *Laminaria saccharina*, oar-weed (× $\frac{1}{12}$). (From Bell and Woodcock, 1971.)

floating life of the plankton (p. 72). The remaining 67 % is effected by terrestrial vegetation.

A more advanced organization is found amongst the lower plants in the Bryophyta, comprising the mosses and liverworts (Fig. 8.5), a comparatively small group of usually rather small organisms (see also p. 163). In appearance they bear some similarity to higher plants, with structures that superficially resemble stems and leaves, but the resemblance is deceptive. Mosses have an erect stem-like axis and are anchored in the soil by branched root-like structures called rhizoids. Liverworts, by contrast, have a leaf-like thallus growing prostrate on the substratum. Both groups require damp habitats, and, lacking true roots and conducting tissue, absorb water over their whole surface. It is this lack of conducting tissue in algae, mosses and liverworts (and also in the fungi, referred to in Chapter 2), that makes a fundamental distinction between these lower (non-vascular) plants and the higher plants (vascular plants, Division Tracheophyta), with their progressive adaptation to a fully terrestrial life.

Flowering Plants

The movement of plant life from water to the more searching mode of life on land required the formation of a vascular (transport) system, together with a skeletal supporting system to compensate for the loss of the buoyancy in water. The consequences are readily seen in the characteristic and familiar form of flowering plants (angiosperms). These are divided into two groups: a larger group (dicotyledons) having embryos (p. 172) with two leaves, and a smaller group (monocotyledons) having embryos with one leaf. All have a rooting system for anchorage, for the uptake of water and dissolved nutrients, and sometimes for storage of reserves, and a shoot, consisting of a stem that bears leaves and flowers. Photosynthesis is largely concentrated in the leaves, the flower is developed out of modified leaves for reproduction, while the stem supports the leaves and the flowers in positions where their functions can best be exercised.

The root, stem, leaves and flowers comprise the organs of the plant, each composed of a number of tissues which exemplify the principle of division of labour amongst the constituent cells of the multicellular body (p. 10), both in plants and in animals. Plant tissues differentiate out of unspecialized cells with thin walls and abundant cell contents, which retain the embryonic capacity for cell division. These cells, called meristematic cells, are initially localized at the growing tips of the root and shoot, where they constitute the apical meristem; the cells differentiated from them form the primary root and stem, and the lateral organs of the stem. It is usual in dicotyledons and gymnosperms (p. 167) for secondary growth and thickening to occur through the activity of lateral meristems. These comprise the vascular cambium, which gives rise to secondary vascular tissues, and the cork cambium, which gives rise to the periderm. The latter replaces the original body covering provided by the epidermis.

A root of a flowering plant (Fig. 3.2a) is composed of three main components, the epidermis, cortex, and vascular cylinder, which differentiate immediately behind the protective root cap. The epidermis is typically a single layer of cells, many of which give rise to the delicate projections called root hairs; these aid absorption, by increasing the surface area of the root and by making close contact with soil particles, and they probably also aid anchorage. They may be covered by a thin cuticle. The cortex consists mainly of parenchyma, composed of the relatively unspecialized parenchymatous cells which form ground tissue throughout the plant, its innermost zone constituting the pericycle. A layer of cells external to this is differentiated in roots as the endodermis,

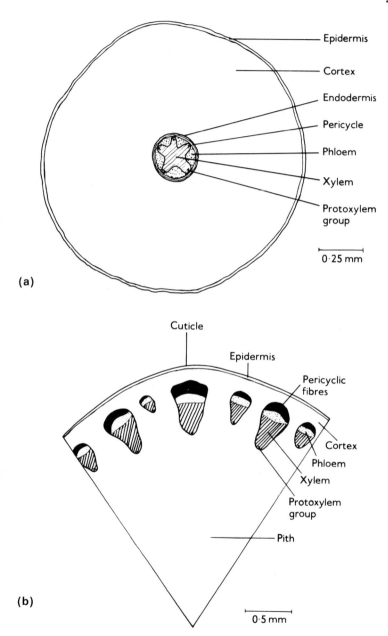

(a)

(b)

Fig. 3.2 Transverse section of (**a**) a young root and (**b**) a segment of a young stem of a generalized flowering plant (From Bell and Woodcock, 1971.)

characterized by some peculiarities of enzyme content suggestive of special physiological functions. Its cells carry a band of suberin (Casparian strip) which is involved in the movement of materials in the roots. Internally to the pericycle lie the conducting strands of the two vascular tissues, xylem and phloem, which we consider later in more detail. The central core of the root may be formed of pith; this is composed of parenchyma or sclerenchyma, the latter tissue consisting of lignified supporting cells which have lost their contents. Alternatively, the core may be formed of xylem. In either case, the main body of the xylem is arranged in a cross-sectional pattern of triangular rays, with protoxylem, the earliest to differentiate, at their periphery. The phloem lies between these rays, so that the whole vascular system forms a symmetrical pattern.

The stem of flowering plants (Fig. 3.2b) is covered by a single-layered epidermis, which is often protected by a cuticle (p. 14). Stomata may be present, in plants in which the stem carries out photosynthesis, but these are more characteristic of leaves, and will be referred to again later. Beneath the epidermis is a ground tissue, containing parenchymatous cells, and also elongated collenchymatous ones, the latter having thicker walls and contributing to support. Within the ground tissue is the vascular tissue. In dicotyledons, this typically forms a cylinder, with a symmetry differing from that of the root in that the xylem and phloem lie along the same radius, with the phloem lying externally to the xylem. The two tissues may form separate vascular bundles, sometimes arranged as a continuous ring, but commonly separated by parenchyma of the ground tissue, which also forms pith in the central axis of the stem. In monocotyledons the vascular bundles are usually scattered irregularly instead of forming a cylinder, but the principle is the same, with the phloem on the outside. Thus the mechanical strength of a plant is based on either a cylindrical axis or on a reinforcing rod, exemplifying principles that are also exploited in animal skeletons, as well as in the designs of construction engineers.

The leaves, with their large surface area, are adapted to their major importance in photosynthesis (Fig. 3.3). The surface is protected by epidermal cells, usually arranged as a single layer, and providing scope for adaptation of the plant to various types of environmental hazard and stress. The cells responsible for photosynthesis constitute the mesophyll, which is differentiated into palisade and spongy mesophyll tissues. The palisade cells, which lie more peripherally, are densely packed, and are elongated at right angles to the transverse plane of the leaf, while the cells of the spongy mesophyll are loosely packed, irregular in shape, and separated by extensive air spaces. Both types of tissue have abundant chloroplasts. Leaves arise at swellings on stems called nodes, increase

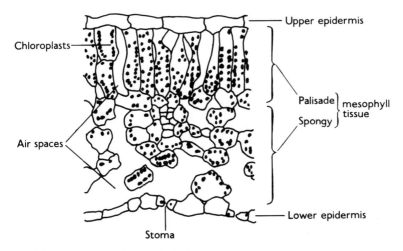

Chloroplasts

Air spaces

Upper epidermis

Palisade ⎤ mesophyll
Spongy ⎦ tissue

Lower epidermis

Stoma

Fig. 3.3 Transverse section of a leaf of *Impatiens parviflora*. (From Bell and Woodcock, 1971.)

in length of the stem resulting from elongation of the internodes. Buds, protected by scales, are leaf primordia and internodes that have not yet lengthened.

The carbon dioxide required for photosynthesis is taken up from the atmosphere, entering the leaf through openings called stomata, of which there are some 30 400 per cm^2 on the lower surface of the leaves of apple trees, for example. Each stoma is surrounded by a pair of guard cells, with which are associated modified epidermal cells (subsidiary cells). The guard cells are elongated, often somewhat kidney-shaped, with localized thickenings of their inner walls. When they are fully turgid (Fig. 3.4), their outer walls tend to swell outwards, carrying the less extensible inner walls after them; this opens the stomata. When the guard cells become flaccid, as happens if the plant is deprived of water, they relax into a position which closes the stomata, a reaction which provides some safeguards against undue water loss. The risk of this occurring is a consequence of the process called transpiration, which is the movement of water up from the roots to the leaves, and its loss from these by evaporation. Whether transpiration is of direct benefit to the plant is not entirely clear. It has been suggested that it is simply an inescapable consequence of photosynthesis, but, as we shall see, it may be an essential factor in promoting the movement of water and mineral salts through the plant.

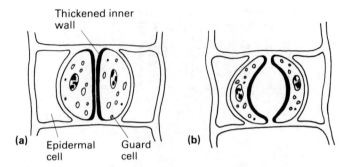

Thickened inner
wall

(a) Epidermal   Guard   (b)
cell   cell

Fig. 3.4 Diagrams of stomata in surface view. (**a**) Stoma closed, with guard cells flaccid. (**b**) Stoma open, with guard cells under turgor pressure which forces their thin outer walls outwards.

The Nitrogen Cycle

Photosynthesis, fundamentally important though it is for the maintenance of plant and animal life, provides, of course, for only part of the nutritional requirements of the green plant. We have outlined earlier the range of elements required. Outstanding amongst these is nitrogen, which, because it is required for the synthesis of proteins, has a particularly marked effect on plant growth. It is an abundant element, providing about four-fifths of the atmosphere, but is present there in a biologically inactive form, and little of this enters directly into plant tissues. Most plants take up their nitrogen from the soil, where it becomes available to them through the operation of a complex 'nitrogen cycle' (Fig. 3.5). Some of this soil nitrogen is derived from the biological breakdown of the waste products and dead bodies of plants and animals (p. 43), but its ultimate source is the atmosphere. Small amounts of nitrogen compounds are formed there by electrical discharge, and are washed into the soil with the rain. This, however, accounts for only a small uptake, perhaps no more than 0.9 kg per hectare per year. By far the largest contribution comes from the biological fixation of nitrogen, in which organic nitrogen compounds are formed from atmospheric nitrogen by living organisms, which are estimated to take up in this way about 100 million tonnes of nitrogen per year. However, our agricultural activities, which result in erosion and leaching as well as harvesting, have made it necessary for us to supplement this contribution by inserting nitrogenous fertilizers into the cycle. These are now derived almost entirely from the industrial fixation of nitrogen by the Haber–Bosch catalytic process. Developed shortly before World War I, this process, which requires the use of

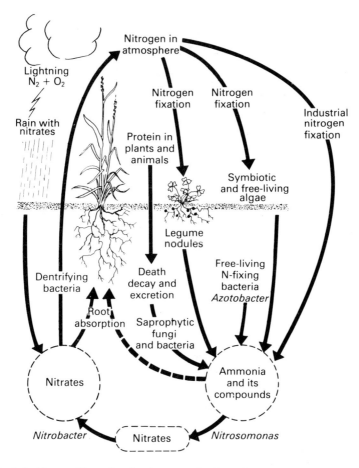

Fig. 3.5 Diagram illustrating the nitrogen cycle. (From Stevenson, 1970.)

our diminishing reserves of fossil fuels, proved also to be a timely source of military explosives. It is estimated to provide at present about one-quarter of the total input of nitrogen into the global cycle, which itself is of the order of 2.4×10^8 tonnes per annum.

Nitrogen fixation

Green plants do not themselves carry out nitrogen fixation, the organisms responsible being blue-green algae and bacteria. Many species of blue-green algae can fix nitrogen, either as free-living

54

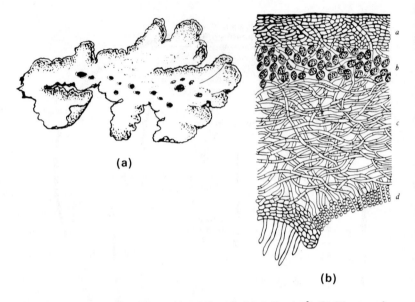

(a)

(b)

Fig. 3.6 (a) *Sticta* sp., a lichen with a foliose (leafy) thallus ($\times \frac{2}{3}$). (b) Diagram of a section of a foliose lichen. The fungal hyphae form the dense upper cortex (1) and the medulla (3); the algal cells (gonidia, 2) lie between these. 4, part of one of the cups (cyphellae) which may serve for aeration. (From Stevenson, 1970.)

organisms or in lichens, which are associations of algae and fungi, to which other types of algae also contribute. In lichens the relationship is so specialized that the plant body (thallus) looks like a single organism, although microscopic study shows it to be composed of fungal hyphae associated with a thin layer of algae (Fig. 3.6). These associations are examples of symbiosis, which is the intimate and prolonged association of two different species of organism. Symbiosis takes many forms, differing in the degree of specialization involved. Parasitism is one example (p. 45), but the components of lichens have a more balanced relationship than that. It is supposed that in this instance the algae benefit from protection and from the uptake of fungal metabolites, although against this it is thought that they may suffer by the loss of nutrients to the fungus, which is said to weaken the algal cell walls. As for the fungus, this certainly benefits from the uptake of nutrients from the algae, for it is unable to synthesize these itself, and it is doubtful whether it can be said to suffer any disadvantage. Probably, therefore, this particular association can best be regarded as mutualism, which is

defined as a symbiotic relationship in which both partners derive
benefit, and which is essential for at least one of them. But whatever
the formal definition preferred, it is evident, from the ubiquity of
lichens on rocks, stones and tree trunks, and from the estimate that
some 15 000 species of fungi contribute to them, that this association
is a highly successful one. It flourishes even in the cold of the Arctic
tundra, where, as reindeer moss (actually the lichen *Cladonia
rangiferina*), it forms the main component of the vegetation.
Nevertheless, it is particularly sensitive to human distortions of the
environment, its disappearance from the scene being a conspicuous
index of pollution by sulphur dioxide. So also is the spread of the
dark (*carbonaria*) variety of the peppered moth mentioned in
Chapter 1. It replaces (although not entirely) the light-coloured
variety which is adapted to match the colour of lichens on the bark
of trees. However, in addition to protective colouration (crypsis),
physiological characteristics, including hardiness and growth rate, are
also involved in this now classic example of the adaptive adjustment
of polymorphism (p. 19) through the action of natural selection.

Despite the environmental importance of lichens, the main agents
of nitrogen fixation are members of another symbiotic relationship;
this is formed by certain genera of bacteria which live in intimate
association with plant roots. The plants mainly concerned are
legumes (Papilionaceae), but there are a few other species as well,
one being the alder tree (*Alder glutinosa*).

The symbionts of legumes (which may be regarded as a special
case of the wide range of micro-organisms that are closely associated
with plant roots, pathogenic as well as symbiotic) are bacteria of the
genus *Rhizobium*. As so often in symbiotic relationships, the partners
are brought together by chemical communication, a phenomenon of
widespread importance in ecological relationships (p. 222), which here
makes possible a highly specific mutual recognition. The roots of the
legume release a lectin which binds to carbohydrate groups on the
rhizobia (Fig. 3.7). A secretion from the bacteria causes the root hairs
to become curled, and the bacteria are then able to enter the root by
penetrating these hairs, passing on to invade the parenchyma cells,
where they transform into swollen forms called bacteroids. The cells
become filled with dense masses of these, but seem not to be
adversely affected; a good indication of the close mutual adaptation
involved. Other cells react to the invasion by dividing and forming
conspicuous swellings or nodules; the bacteroid-containing cells lie in
the centre of these, closely associated with vascular tissue continuous
with that of the host.

Within the nodules the bacteroids fix nitrogen and incorporate it
into amino acids by a pathway which has been traced by the use of
radioactive nitrogen ($^{15}N_2$). The first step is the reduction of

56

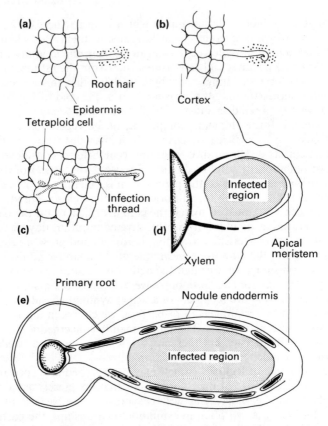

Fig. 3.7 Initiation and structure of pea nodules. (**a**) Rhizobial bacteria aggregate round root hairs. (**b**) Root hairs curl. (**c**) Bacteria infect root hair and move through the root hair and inner cortex; this stimulates meristematic activity. Polyploidy is perhaps induced by the advancing infection thread. (**d**) A central infected region and an apical meristem become distinguishable. (**e**) Longitudinal section through nodule showing central infected region, apical meristem and nodule endodermis. (After Stewart (1966). *Nitrogen Fixation in Plants*, Athlone Press, London.)

molecular nitrogen to ammonia by the enzyme nitrogenase. The ammonia is converted into glutamine through the agency of glutamine synthetase, glutamate being then formed in the presence of glutamate synthetase. From glutamate, which is an important store of nitrogen in both plants and animals, other amino acids can readily be derived by transamination: the transfer of amino groups through the action of enzymes called transaminases. The production of alanine is an example:

$$
\begin{array}{llll}
\text{COOH} & \text{COOH} & \text{COOH} & \text{COOH} \\
| & | & | & | \\
\text{CH.NH}_2 \;+ & \text{CO} \;\;\rightleftharpoons & \text{CO} \;\;+ & \text{CH.NH}_2 \\
| & | & | & | \\
\text{CH}_2 & \text{CH}_3 & \text{CH}_2 & \text{CH}_3 \\
| & & | & \\
\text{CH}_2.\text{COOH} & & \text{CH}_2.\text{COOH} &
\end{array}
$$

Glutamate   Pyruvate   α-Ketoglutarate Alanine

This form of nitrogen fixation is completely dependent upon the symbiotic association; the legumes cannot fix nitrogen in the absence of the bacteria, nor can the latter do so when they are cultured outside the plants.

In addition to the transformation of the bacteria into bacteroids, another condition for nitrogen fixation is the production within the plant of a red pigment, leghaemoglobin, which can only be formed in the presence of the bacteria. This pigment, which is chemically similar to the myoglobins of animals, and which, like them, binds oxygen reversibly (p. 102), aids the passage to the bacteroids of the oxygen required by them for the formation of the ATP which drives the reduction of the nitrogen. Further, nitrogenase is inhibited by oxygen, but is able to function in the bacteroids because leghaemoglobin, which has a remarkably high oxygen affinity, lowers the concentration of free oxygen around them. These subtle reactions show what a high level of adaptation has been reached in root nodules. The association may be regarded as another example of mutualism, with advantages accruing to both partners. The plant, however, takes most of the fixed nitrogen, and it is possible that the bacteroids are entirely dependent on their host for their own supply, being unable to assimilate themselves the nitrogen which they have fixed.

No less subtle is an adaptation found in certain of the nitrogen-fixing blue-green algae already mentioned. Some of these fix nitrogen anaerobically, but others, despite their dependence on the oxygen-sensitive nitrogenase, achieve aerobic fixation. Here the problem is solved by the nitrogenase of the aerobic forms being localized within specialized cells (heterocysts) which provide an anaerobic environment. These cells maintain an essentially mutualistic relationship with neighbouring vegetative cells, providing them with combined nitrogen and receiving in exchange the organic substrates needed for the fixation process.

Nitrogen fixation is also effected by free-living soil bacteria, which include both anaerobic forms (e.g. *Clostridium*) and aerobic ones (e.g. *Azotobacter*). How much these bacteria contribute to the total fixation of atmospheric nitrogen is uncertain, but it seems likely that they are of only minor importance. One reason for supposing this is that the nitrogen content of naturally fertile soils is likely to reach a level capable of suppressing the fixation of nitrogen by *Azotobacter*. Another reason is that the anaerobic conditions favouring *Clostridium* are unsuitable for plant roots, so that its nitrogenous products are more likely to be lost by drainage than to be taken up by plants.

Decomposers

Free-living bacteria have another and much more important role in the soil than the fixation of nitrogen. They are the most abundant members of a vast complex of soil micro-organisms, which amount to as many as 5000 millions or more per gram, living in the air spaces and the films of water that surround the soil particles. Others are the filamentous bacteria (Actinomycetes), protozoans, slime moulds (Myxomycetes), fungi and algae. Associated with these are a diversity of animals, including earthworms, mites, ants and other insects, centipedes, millipedes, woodlice, slugs and snails. These, too, are present in enormous numbers, grassland soil, for example, containing some 20 arthropods per $10\,cm^3$. The members of this complex function ecologically as decomposers, forming an essential link between the autotrophic producers and the heterotrophic consumers, although to some extent overlapping the latter. They ensure the continuous breakdown of the leaf litter and other organic material which collects on the soil (p. 43), and in so doing help also to complete the biological nitrogen cycle, only part of which has so far been described.

Plant litter is rich in carbohydrate, with a low ratio of nitrogen to carbon. Its carbohydrate component is broken down to CO_2 and water by bacteria and fungi which digest the cellulose, aided in this by members of the soil fauna which break the material down mechanically, and by earthworms, which pull fallen leaves into the soil. The CO_2 escapes into the air, becoming available again for photosynthesis, and leaving the material called humus, which has a high nitrogen content. The protein in this is broken down by bacteria and other organisms, such as nematode worms, which feed saprophytically (p. 45) upon it. Some of the products are incorporated into their own bodies, but there is a considerable surplus which enters the soil in solution. This could, in theory, be used by plants, but in practice it is largely exploited by chemotrophic micro-organisms (p. 32). These operate in a chain, the first link in which is exemplified by the bacterium *Nitrosomonas*, which obtains its energy by oxidizing ammonia to nitrite:

$$NH_4^+ \rightarrow NO_2^-$$

The next stage is exemplified by *Nitrobacter*, which oxidizes nitrite to nitrate:

$$NO_2^- \rightarrow NO_3^-$$

With the energy so obtained, these bacteria are able to assimilate CO_2 from bicarbonate ions (HCO_3^-) and to build up glycogen and other reserves as a basis for respiratory metabolism and growth. These processes, which constitute what is called nitrification, leave a surplus of nitrate which is absorbed in solution into the roots of green plants and provide their main source of nitrogen, apart from legume nodules.

There is also a reverse process, called denitrification, effected by bacteria which operate both aerobically and anaerobically to break down nitrite or nitrate, using these as hydrogen acceptors and releasing gaseous nitrogen or nitrous oxide. The resulting loss of molecular nitrogen from the soil is estimated to be of the order of 3.7 kg of nitrogen per hectare per year. However, these organisms play a useful part in eutrophic lakes, which are lakes that accumulate quantities of nitrogen and other nutrients from inflows of sewage and fertilizers. Eutrophication results in excessive algal growths which can lead to a reduced oxygen content owing to bacterial and fungal decomposition. Denitrifying bacteria, because of their ability to effect the anaerobic breakdown of nitrates, can help to correct this man-made disturbance.

The nitrogen cycle (Fig. 3.5), vital to the development, growth and maintenance of plant and animal life, is accompanied by corresponding cycles of other elements. For example, we have seen (p. 38) that hydrogen sulphide, released from dead organic matter by heterotrophic decomposers, is oxidized by sulphur bacteria to sulphate. This is taken up in solution by autotrophic plants, together with phosphate and cations such as potassium, magnesium and calcium, all following their own cycles, and to be eventually returned again into the soil, either directly from the plants, or indirectly from the animals that have exploited them. Always, through wastage and death, there is abundant material for recycling within the organic interrelationships of ecosystems.

Uptake of Nutrients

The nature of the soil has a major bearing upon root structure, and, indeed upon the whole of uptake, for most of the water taken up by rooted plants enters through their root systems. In general, dry soils promote the development of deep and much branched roots, whereas wet soils, with poor aeration, encourage shallow roots. As always, however, there are exceptions to these generalizations; cacti, for example, may have shallow roots, which permit rapid uptake of water entering the soil after occasional light rains.

Most absorption takes place through the younger parts of the roots, just behind the growing apex, for here the epidermis is thin-walled, and root hairs are maximally developed. The total absorptive area available is impressively large. Thus the total surface area of the root system, including that of the root hairs, of a 4-month-old plant of rye (*Secale cereale*) is 22 times the total surface area of the shoot, including the area of the cells bordering the internal air spaces as well as the external surface area. Part of the reason for this is that plants are always forming new root tissue at the growing points, for the older surfaces tend to become impermeable, and most absorption must therefore take place through the younger parts. This growth is also needed to permit the exploitation of further sources of soil water and nutrients; to some extent it is the equivalent of the movements of animals.

Initially the water enters the cytoplasm of the epidermal cells, and does so passively, along what is termed the water potential gradient. Osmotic mechanisms are a major factor, including the turgor pressure of the cells and their solute concentrations. Whether active uptake plays any part is uncertain, but it seems unlikely to do so. The water must then pass to the vascular system, but the path that it follows is uncertain. It may be that it moves inwards through the plasmodesmata that connect the cells into what is virtually a continuous cytoplasmic system, but it is also possible that there may be other paths; the cells walls, for example, or the extracellular air spaces.

Ions are taken up into the root from solution in the soil water, moving into and through the root across plasma membranes. This process is a selective one, taking place against concentration and electrochemical gradients, and involving, therefore, some form of active transport (p. 13). The path followed by the ions is not entirely clear, but one view is that they first accumulate within the cytoplasm of the epidermal cells and then diffuse along the plasmodesmata, perhaps aided by cytoplasmic streaming. As with the supply of water, the effectiveness of any particular soil depends upon many factors. Amongst these are its chemical and physical composition, the related composition of the soil solution, the availability of micro-organisms to contribute to the nutritive cycles, and aeration adequate for the supplying of oxygen and the removal of respiratory carbon dioxide.

Symbiosis may play a part in nutrition, additional to those aspects of it mentioned earlier, for root absorption can be aided, and especially in poor soils, by mycorrhizas. These are symbiotic non-pathogenic associations between fungi and the roots of higher plants, including trees, orchids and heathers. External (ectotrophic) mycorrhizas (ectomycorrhizas) are found on forest trees such as beech and pine, the feeding rootlets being enclosed by a layer of fungal tissue from which hyphae extend into the interior of the root and, to a less extent, into the soil (Fig. 3.8). Experiments have shown that the fungal tissue has a high rate of nutrient uptake and that it also accumulates reserves, notably of phosphates, which can pass into the host. Against this, the tree has to provide carbohydrate for the fungus, to an extent which may be as much as 10% of the carbohydrate which it uses for timber production. The compensating benefit to the tree lies in the absorptive and storage capacity of the fungus, which is probably of particular value in seasonal climates, enabling the ectomycorrhiza to exploit temporary increases in nutrient supplies.

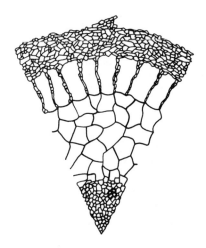

Fig. 3.8 Transverse section of *Fagus* mycorrhizal root showing external fungal sheath, with hyphae passing inwards and, to a slight extent, outwards. (From Clowes, 1951. *New Phytol*, **50.**)

Transport of Nutrients

Transport between root and leaves is effected by the conducting strands of the xylem and phloem, through which move the nutrient fluids that constitute the plant sap. The main path for the passage of water and mineral salts from roots to leaves is the xylem, a complex lignified tissue characteristic of vascular plants. Included in the xylem are unspecialized parenchyma cells, with some storage function; the older ones may become lignified. There are also supporting fibres formed of elongated cells of the type called sclerenchyma. Such plant

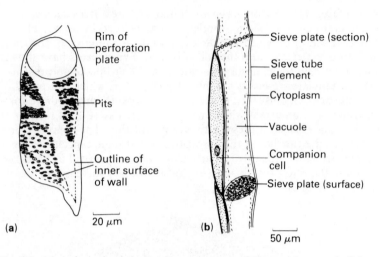

Fig. 3.9 (a) A single vessel element from a macerate of the secondary wood of the ash (*Fraxinus*). (b) Longitudinal section of a sieve tube and companion cell of cucumber (*Cucumis*). (From Bell and Woodcock, 1971.)

fibres are typically lignified, and lack living contents when mature, but xylem fibres may retain their protoplasts for a long time. The conducting elements, which are of two types, called vessel elements (or members) and tracheids, are formed of elongated cells with thick lignified walls (Fig. 3.9a). The vessel elements are arranged as chains of cells, the individual elements being in communication through one or more pores which are located in perforation plates at the ends of the cells, or, less typically, along their sides; pits are also present. These chains may reach a length of several feet. The tracheids are single cells, pointed at their ends, and lacking pores (imperforate), and with their pits closed by membranes.

Xylem sap is a clear liquid, with a low concentration of solutes and with correspondingly low viscosity. Total dry matter rarely exceeds 1 %, and the carbohydrate content, in particular, is almost negligible. The function of the xylem in the transport of this sap is not in doubt, but it is uncertain how the transport is achieved, for there is nothing in the vascular tissue of plants comparable to the contractile tissue that effects transport in animals. If a root is severed, it is often possible to observe exudation of fluid from the vessels, indicating a positive hydrostatic pressure in the xylem sap. However, manometric measurements show that this pressure, which is probably an osmotic phenomenon, is often much too low to be able to force water up to the leaves. The most acceptable explanation of the movement of the sap is embodied in what is termed the transpiration-cohesion-tension theory. According to this, the loss of water from the leaves and shoots by evaporation draws water from the xylem, and this draws a column of water upwards from the roots through the conducting tissue. This presupposes two requirements, both of which can be satisfied.

Energy is required, and this comes from the solar radiation which causes evaporation. Cohesion of the water molecules in the column is also required, but these molecules do indeed have a very high cohesive attraction. Calculations attempting to relate this cohesive force to conditions obtaining within the plant have given somewhat discordant results, but it appears that the tensile strength of the xylem sap would be adequate to permit a column of water being pulled to the top of the tallest trees; no small matter, for these can approach heights of 80 metres and more! Supporting evidence comes from the construction of inorganic models which can be built to satisfy the requirements of the theory.

The phloem, the second component of the conducting tissues, conveys the phloem sap. Studies with fluorescent materials and radioactive tracers show that this moves in both directions, whereas movement in the xylem is mainly upwards. The difference is associated with fundamental differences in the structure of the two types of vascular tissue. Phloem is composed of parenchyma, sclerenchyma fibres, less elongated elements called sclereids, sieve cells (sieve elements), and companion cells. The elongated sieve cells of angiosperms are arranged in longitudinal rows of cells to form columns called sieve tubes (Fig. 3.9b), each cell having specialized pores which form a sieve plate at or near the end of the cell. These sieve cells differ from xylem elements in being living cells which retain their contents, although they are unique amongst plant cells in losing their nuclei as they mature, while their mitochondria also degenerate. Each is associated with a companion cell, which is a normally constituted cell that arises, with its associated phloem element, from a common mother cell. It is likely that it contributes in some way to the functioning of the sieve elements, perhaps facilitating the passage of material out to neighbouring tissues.

Phloem sap has a higher concentration of solutes than has xylem sap, and is correspondingly more viscous. Its composition can be elegantly studied by allowing certain aphids (not all are as obliging) to insert their stylets into the sieve tubes, which they do with great precision, and then cutting away their bodies. The sap, which continues to exude from the cut ends of the mouth parts, can then be collected in a very pure form. It differs from xylem sap in being more alkaline and more viscous, and in its high concentration of solutes, particularly of sugar, the concentration of which ranges from

2–25 % w/v. (We have seen that carbohydrate photosynthesized in the leaf may be stored there as starch, but it is primarily as sucrose that it is transported away in the phloem sap.) Mineral salts, amino acids and other substances are also present. The composition, especially of amino acids, shows marked seasonal variation, providing a signal which has been exploited in the life cycles of aphids (p. 186).

The complex branching of the phloem ducts enables them to reach most parts of the plant, but how movement of the phloem sap is contrived is uncertain, although it is clear that living cells are involved. Probably the most likely explanations are either that solute particles are moved by mass flow of the solvent, or that transport takes place by protoplasmic streaming through the sieve elements and through the cytoplasmic strands which can be identified in the sieve pores. Other suggestions include the possibility of some form of active transport, or of diffusion through molecular films within the cytoplasm of the sieve elements. The diversity of these suggestions is sufficient to show that the problem of phloem transport remains unresolved. This is unfortunate considering the importance of its contribution to the environmental relationships of the plant.

4 Consumption

Ecosystems

The photosynthetic activity of plants, and their anabolic building of carbohydrates, fats and proteins from inorganic precursors, constitutes gross primary production. Some 10–20% of the energy thereby assimilated is used up in respiration; the balance, represented by the accumulation of plant biomass (the weight of plant material), constitutes net primary production. This is succeeded by a multi-tiered system of secondary production, in which animals feed upon plant energy stores and add to their own biomass by molecular re-organization of the food.

This flow and processing of energy depends upon the organization of animals and plants into integrated communities called ecosystems (Fig. 4.1). These suffer continuous loss of energy as heat of metabolism, and through the activities of decomposers, but they are able to maintain themselves, and to grow to optimum size, because this energy can be replaced from the normally limitless supply of solar radiation. Moreover, ecosystems, like other biological systems at all levels of organization, have evolved regulating mechanisms which maintain them in a dynamic steady state. This is the condition called homeostasis, which we shall encounter later in other contexts as well. Here it depends upon the recycling of nutrients within the system and upon the storage of energy in reservoirs, such as forest trees and organic detritus, which can be drawn upon as need arises.

The detailed patterns of ecosystems vary with the nature of the habitat, but the underlying principles remain the same. Plant material is eaten by primary consumers, either directly by grazing herbivores, or less directly by detritus feeders. Primary consumers are eaten by carnivores, and, because of the enormous range of size of animals, there may be larger carnivores eating the smaller ones. Ecosystems thus have a hierarchical organization, with energy flowing from one level, called a nutritional or trophic level, to another. Plants occupy the basic level, primary consumers the second, and secondary consumers the third, with perhaps large carnivores occupying one or more additional levels, culminating in a small number of 'top' carnivores. Organisms at the higher trophic

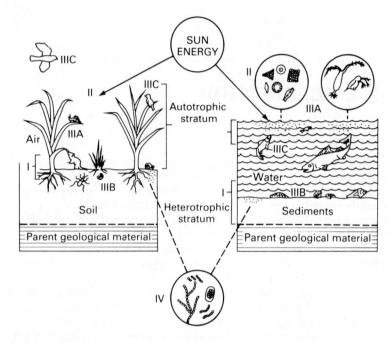

Fig. 4.1 Comparison of the gross structure of a terrestrial ecosystem (a grassland) and an open-water ecosystem (either freshwater or marine). Necessary functional units are: **I**, Abiotic substances (basic inorganic and organic compounds). **II**, Producers (vegetation on land, phytoplankton in water). **III**, Macroconsumers or animals – (A) direct of grazing herbivores (grasshoppers, mice, etc. on land; zooplankton, etc. in water); (B) indirect of detritus-feeding consumers (soil invertebrates on land, bottom invertebrates in water); (C) 'top' carnivores (hawks and large fish). **IV**, Decomposers, bacteria and fungi of decay. (From Odum, 1966.)

levels will in any case be fewer than those at the lower ones, because the transfer of energy is accompanied by its metabolic loss. So within the ecosystem there is a pyramid of numbers, the highest trophic level, with the smallest numbers, occupying the apex of the pyramid.

The concept of trophic levels has been criticized on the ground that particular species may obtain their energy from more than one level. Detritus feeders, for example, may consume animal as well as plant material. It is intended, however, as a broadly functional concept, not as a detailed classification, and within this limit it provides a valuable aid to description and analysis. Nevertheless, the formidable complexities of ecosystems should not be underestimated. Energy flows along food chains, made up of one or

more species at each trophic level (e.g. flowering plant →
caterpillar → bird, or grass → cattle → man), but any one
species in a chain will often feed upon more than one species at the
next lower level, so that in practice the food chains in an ecosystem
are interrelated to form complex trophic systems of food webs (Fig.
4.2).

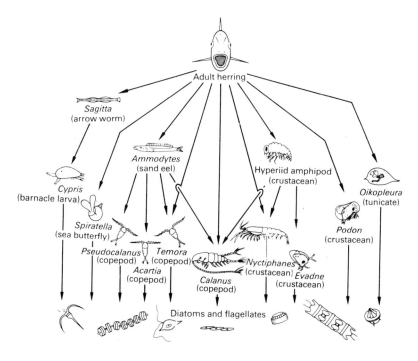

Fig. 4.2 The feeding relationships of adult herring. (After Hardy, A. (1959). *The
Open Sea*, II. Collins, London; from Phillipson, 1966.)

These principles are fundamental to the science of ecology, which
is the study of the relationships of plants and animals with each
other and with every aspect of their environment. They are
considered in detail in another book in this series, but in the present
context they provide a basis for the assessment of the adaptive
organization of individual species, of which we shall now examine a
few illustrations.

Food Requirements of Animals

Energy is transferred from plants to animals in the form of

carbohydrates, fats (or lipids) and proteins, which are their most obvious food requirements. The carbohydrates are predominantly in the form of starch and cellulose. Of these, the starch is readily broken down, in the way that we shall consider shortly, into glucose, which can then be metabolized along the pathways that we have already discussed, or it can be built up into glycogen, which is the principal form of stored carbohydrate in animals. Strangely enough, however, cellulose, despite the enormous quantities available, cannot be directly utilized by many animals, since they do not, with rare exceptions, produce the digestive enzymes required for its breakdown. Nevertheless, this important carbohydrate resource is not wasted, for, as we shall see, the animal world has evolved ingenious devices for making use of it.

Fats, which are esters of glycerol and fatty acids, usually as triglycerides (p. 6), are as readily exploited as is starch, being broken down in part into free fatty acids, but probably in part only to monoglycerides. These are then recombined into the fats characteristic of the particular species.

Proteins are broken down into amino acids, which are then recombined into the proteins required by the species, but here a difficulty arises which is peculiar to this class of material. Some amino acids can be synthesized by animals in various ways; alanine, for example, can be formed from glutamic acid and pyruvic acid by transamination (p. 56). Other amino acids (lysine is an example) cannot be synthesized, so, because of this limitation, animals must have certain amino acids provided for them in their food; these are termed essential amino acids. Ten are required by the human infant, and it is a remarkable indication of the uniformity of biochemical organization, mentioned earlier, that the growing rat, the chicken, the house fly (*Musca*) and a protozoan, *Tetrahymena*, have these same requirements: arginine, histidine, isoleucine, leucine, lysine, methionine, threonine, tryptophan, phenylalanine and valine. Autotrophs can, of course, synthesize all of the twenty common amino acids; some of this capacity must presumably have been lost very early in animal evolution.

These amino acids are not the only requirements essential for animal life. There are also a number of vitamins, which are organic substances of low molecular weight, that the animal cannot make for itself, although they are essential for its normal functioning. Some are fat-soluble, some water-soluble; some are virtually universal requirements throughout the animal kingdom, others are needed only by particular groups.

One of the water-soluble vitamins, vitamin B_1 (thiamine), will serve to illustrate the kind of role played by several of these substances. We have noted that many enzymes can only function in conjunction with non-protein coenzymes or prosthetic groups. During the oxidative decarboxylation of pyruvic acid, which results in the formation of acetyl-CoA, the pyruvic acid combines with thiamine pyrophosphate (cocarboxylase) to form acetyl-thiamine pyrophosphate, with the release of carbon dioxide. The acetyl group is then transferred to lipoic acid, and from there to coenzyme A. In this chain of reactions (cf. p. 39) the thiamine pyrophosphate and the lipoic acid function as coenzymes. Other water-soluble vitamins of the B complex are also essential coenzymes, and, like vitamin B_1, participate in reactions which are fundamental to living organisms, both plant and animal. In general, these vitamins are obtained by animals, directly or indirectly, from plant sources. The inability of animals to make them for themselves, just like their limited capacity for synthesizing amino acids, is a result of some biochemical degeneration; indeed, the same might be said of the very origin of animals, which is founded on the loss of the power of photosynthesis.

However, not all the vitamins are obtained by animals from plants. The fat-soluble D vitamins (D_2, or calciferol, and D_3), which aid the absorption of calcium from the intestine of vertebrates, and are also involved in the regulation of calcium and phosphate levels in the blood, are a case in point, for these vitamins, required only by vertebrates, are synthesized by them from cholesterol. Vitamin A (A_1 and A_2), also fat-soluble, are precursors of the light-sensitive pigments of the vertebrate eye, and are probably needed, again only by vertebrates, for the normal development of their epithelial tissues. Here plants do make a contribution, providing carotenoid pigments (widely distributed in photoreceptor systems from micro-organisms upwards), from which vertebrates can synthesize the vitamin. So also can crustaceans, which makes it possible for fish to obtain much of their supply from those animals. It would seem that these fat-soluble vitamins are a product of the further evolution in the vertebrates of molecules that were already widely distributed.

Animals, like plants, require also a variety of inorganic substances, either as macronutrients or as micronutrients. Included among the macronutrients are calcium, iron, magnesium, potassium and sodium. We shall see reasons for the need for some of these later. Amongst the micronutrients are cobalt, copper, manganese, zinc, iodine and fluorine. Some of these nutrients are essential components of enzymes (it will be recalled that the cytochromes are iron-porphyrin compounds), but their functions are diverse and vary from group to group to an extent which makes an understanding of their functions vital for man's success in animal production.

Iodine, for example, is a characteristically vertebrate need, being a component of the two hormones secreted by the thyroid gland.

Iodine deprivation leads to underproduction of these hormones and, because of this, to stunted growth and defective development of the nervous system. Cobalt, which is a component of vitamin B_{12}, is required in relatively large amounts by ruminant mammals because they depend upon the bacterial flora of their rumen (p. 92) for the production of that vitamin, while copper is needed by sheep to ensure the normal structure of their wool.

Marine Ecosystems

The relation between primary producers and consumers, although always conforming to the general pattern outlined above, differs fundamentally in the details of its organization according to whether the primary production takes place in water or on land (Fig. 4.1). Thus the oceans, covering some two-thirds of the earth's surface, are the basis of an ecosystem which is outstanding in its size and stability. Conditions here are highly favourable for life (cf. pp. 112, 128), which is one of the reasons why it is thought to have originated in primeval oceans.

Marine vegetation is most conspicuous on rocky shores, where seaweeds can secure attachment, and where they are exploited as food by animals and by man for various purposes (the Pacific *Gelidium* is the source of agar). Molluscs are outstanding exploiters of this food source, making use of their radula, a ribbon-like structure supported within a radula sac by a firm rod (the odontophore), and bearing horny rasping teeth (Fig. 4.3). This organ, very characteristic of the phylum, is used by the chitons (Fig. 4.4a) in what is doubtless the primitive way, rasping small particles from encrusting algal growths, or from the fronds of larger

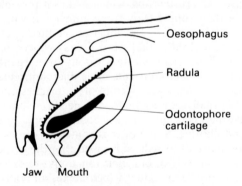

Fig. 4.3 Diagrammatic vertical sections through the buccal mass of the gastropod mollusc, *Lymnaea stagnalis*, to show the radula and associated structures during feeding.

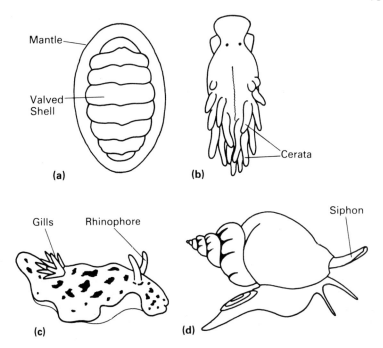

Fig. 4.4 (**a**) Diagram of a typical chiton, dorsal view, showing the 8-valved shell surrounded by the body wall (mantle). (**b**) *Alderia*, dorsal view; the cerata are respiratory outgrowths. (**c**) *Glossodoris*, a sea slug. (**d**) *Buccinum*, lateral view.

specimens. The habit has lent itself to extensive diversification, well seen in the class Gastropoda. Members of the order Saccoglossa use their radula for extracting the cell contents of green algae, in a relationship of great elegance, the algal filament being pierced by one tooth, and the size of this being precisely matched with the size of the algal cell. Thus *Alderia* (Fig. 4.4b), a brackish form, has a tooth 33–35 μm in diameter, adapted for piercing filaments of *Vaucheria*, which are 40–60 μm in diameter, and which it can empty at the rate of ten per minute.

The versatility of the radula is seen in its exploitation by molluscan carnivores. The sea slugs (Fig. 4.4c), marine gastropods with secondary gills around the posterior anus, use it to break down the sponge colonies over which they slowly creep. The whelk, *Buccinum* (Fig. 4.4d), bores into the shell of recently dead molluscs. Having detected its food chemically, by drawing water through a siphon and over a sense organ at the base of this, it uses an extensible proboscis to suck out the contents of the body. Other gastropods use the radula to pierce individual members of polyzoan colonies; they then extract the contents with a sucking pump formed by the buccal mass.

The exploitation of algae by molluscs illustrates the general principle that no feeding niche is ever left unoccupied. Nevertheless, the total contribution of shore-living algae to primary production is negligible. Most of this comes from the microscopic plant life living in the plankton, a great assemblage of floating and drifting (but not necessarily passive) organisms, both plant (phytoplankton) and animal (zooplankton), unicellular members of which were probably the sources of much of our petroleum deposits. This contrasts with the life of the sea bottom (benthos) and with those organisms (nekton) which can swim actively and move independently of currents, such as fish and squids.

The solar energy required for primary production in the plankton is rapidly absorbed by water. Blue light penetrates furthest, but 50 % of the energy is at the infra-red end of the spectrum, and nearly all of this is absorbed within the uppermost metre. Unfortunately, chlorophyll absorbs maximally at the red end of the spectrum, but its action is aided by additional pigments which, as we have seen earlier, absorb light at other wave-lengths and pass its energy on to the chlorophylls. Even so, the euphotic zone, in which photosynthesis is confined, is restricted at most to the upper 200 m of the sea, even in the clearest waters, and it is within this upper layer that the active plant life is confined.

Several categories of organism are included in this phytoplankton. The principal primary producers, in fresh water as well as in the sea, are the widely distributed and abundant algae called diatoms (Fig. 4.5). They are characterized by their beautifully sculptured siliceous walls, which, because of their resistance to decay, accumulate as 'diatomaceous earth' at the bottom of seas and lakes. A second group, more abundant in warmer waters, comprises the dinoflagellates (peridinians), which, because they possess two flagella lying in grooves, one transverse and one longitudinal (Fig. 4.5), have some power of independent movement. These are algae (Dinophyceae) which are not easily separated from heterotrophic flagellate protozoans, and are commonly classified by zoologists as flagellate protozoans within the class Mastigophorea; they are then termed Dinoflagellida. The taxonomic complication is a consequence of the mode of evolution of heterotrophic animals from autotrophic plants. This seems to have taken place along more than one line, and the close interweaving of plant and animal forms within the Mastigophorea is a consequence of this, for it is within this group that the transitions must have occurred. Finally, there are multitudes of very minute organisms, measuring less than 5 μm,

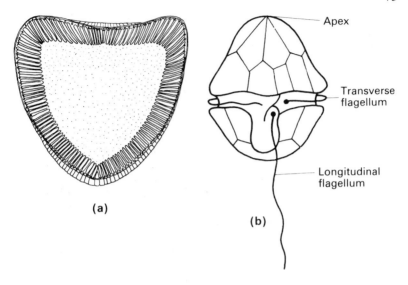

Fig. 4.5 (a) A marine diatom, *Campylodiscus*, × 450. (From Bell and Woodcock, 1971.) (b) *Glenodinium*, a freshwater dinoflagellid with a covering 'armour' of cellulose plates.

sometimes referred to as the nanoplankton (*nanos*, dwarf); mostly flagellates, together with bacteria.

The restriction of primary production to a relatively narrow euphotic zone immediately raises a question. How is the captured energy economically transferred to animal life, which occurs not only in the surface waters but down to the greatest depths (for hardly anywhere in the oceans is some form of life totally lacking)?

One factor is the common occurrence in the zooplankton of vertical migration, made possible because many members of it are powerful swimmers in relation to their small size. Those relying upon cilia (many invertebrate larvae, for example) are necessarily weak in this respect, but fish larvae and crustaceans move with the aid of muscle, the latter group, thanks to their locomotor appendages, attaining speeds in upward movement ranging from 10 to as much as 400 m per hour. Downward movement is, of course, aided by gravity. The vertical range of this migration varies from species to species, being governed especially by sensitivity to light, each species tending to congregate at a particular light intensity. In consequence, the migration commonly shows a regular 24-hour (diurnal) cycle, but this is rarely a simple one, for the animals' responses can also be influenced by other factors, such as cloud cover, disturbance of the surface waters by storms, moonlight, and seasonal fluctuations of light intensity. Presumably this migration is of benefit to the animals. It must protect them from exposure to harmful light intensities, and perhaps the lower rate of metabolism at the lower temperatures of deeper waters provide some economy in the use of energy. But there is also benefit to the whole marine ecosystem, for these vertical movements contribute to the distribution of the energy captured by the phytoplankton.

There is, however, another problem arising from the restriction of photosynthesis to the surface waters. The organisms of the plankton die, and so do the animals that feed upon them. In consequence there is a steady fall to the ocean floor of dead material and animal faeces, together with the dissolved nutrients produced by their bacterial decay. The sediment thereby formed certainly contributes to the life of the benthos, but the dissolved nutrients are not, in general, taken up by animals. (One exception to this is the deep-sea phylum Pogonophora, a group of strange animals that are thought to be related to annelid worms. Pogonophorans have lost all trace of mouth and alimentary tract, and absorb nutrients through their body surface, but they are the exception that emphasizes the rule.) It follows, if this were all, that the dissolved nutrients in the ocean depths would form a sink, unavailable for most animal life or for the phytoplankton of the photic zone. The perpetuation of this sink would rapidly lead to the end of oceanic life.

Escape from this impasse is made possible by a process which well illustrates the close dependence of the life of our planet upon the physical conditions provided. The prevailing winds, in conjunction with the rotation of the earth, bring about vertical water movements which occur especially as upwellings along the western shores of the continents. It is these movements, slow though they are in comparison with the more familiar horizontal currents (up to 10 m per day, as compared with 100 km per day for the latter), which restore the abyssal nutrients to the surface, and thus make a crucially important contribution to the maintenance of marine life.

Microphagy

An essential factor in the development of marine ecosystems (and this applies to freshwater ones as well) has been the evolution by many animals, and not only members of the zooplankton, of a capacity to capture and ingest organisms and particles of microscopic size. This type of feeding, called microphagy, is to be contrasted with macrophagy, which is feeding upon food that is large relative to the size of the animal taking it. The two modes, which are associated in general with two different trophic levels, make possible various types of food chain. Thus microphagous crustaceans, which are dominant amongst the zooplankton, can ingest the phytoplankton and are then eaten by macrophagous cephalopods or by larval and adult fish. Typical chains could be ended by the tuna eating smaller fish which had fed upon the zooplankton, or by the predation of birds, seals and polar bears upon fish. One exceptional case, illustrating the flexibility of ecological adaptation, is provided by the whalebone whales (Mystacoceti). These enormous mammals feed directly upon the krill (euphausiid crustaceans) which form part of the zooplankton, establishing thereby a food chain of impressive simplicity.

A difficulty inherent in microphagy is that individual particles cannot usually be captured separately, but must be swept to the predator in bulk. Because of this, microphagous animals are often filter feeders, equipped with various devices for filtering the particles out of large volumes of water, and often also with devices for setting up water movement, although some filter feeders rely upon the natural movements of water currents. Adaptations for such filter feeding are diverse, ingenious and beautiful, and merit close attention for their intrinsic interest, but only a few examples can be mentioned here.

Dominant amongst the filter feeders of the marine zooplankton are the smaller crustaceans, which are well fitted for this mode of feeding because of the jointed appendages which are one of the characteristics of arthropods. Some of these appendages, by a useful economy of energy, can contribute both to locomotion and to the setting up of food currents, while the hair-like setae, which are often richly developed along their edges, provide for filtering the water and for trapping its contents. The method is well seen in the Copepoda, a group of crustaceans, commonly from one to several mm in length, which dominate in the marine zooplankton.

Calanus is a familiar and much-studied copepod (Fig. 4.6). Appendages on the head (antennae, palps of the mandibles, and the maxillules) maintain a rapid vibration which sets up a large vortex on each side of the body, these being supplemented by smaller vortices set up by other appendages. The water is drawn towards the ventral surface of the body, which, in conjunction with parts of certain limbs, forms a suction chamber. Food, filtered out of the water by setae on the maxillae, is passed forwards to the mouth by setae on other appendages.

Copepods are usually omnivores, taking both phytoplankton and the smaller zooplankton; they thus exemplify feeding at two trophic levels. Another crustacean group, the Cladocera (water fleas), an important component of both marine and freshwater plankton, is predominantly herbivorous. Cladocerans, like copepods, make use of limb movement, suction chamber and filtering setae, but with much

76

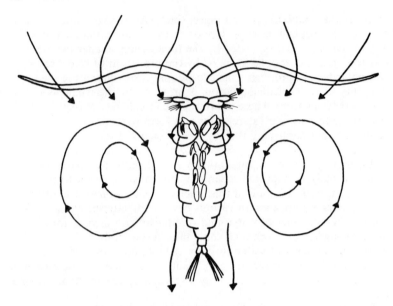

Fig. 4.6 Diagram of ventral view of *Calanus* slowly swimming, to show water currents.

difference in detail, for their limbs are reduced in number, while their bivalved carapace contributes to the suction chamber (Fig. 4.7).

Elsewhere in the animal kingdom it is common for microphagy to depend upon the beat of the cytoplasmic threads called cilia, which are borne at the surface of specialized ciliated cells. These are almost entirely absent from arthropods. This type of microphagy is well seen in the bivalve molluscs, which are typically benthic animals, mostly feeding through the agency of elongated gills (ctenidia) situated within an enfolding mantle cavity (Fig. 4.8). These gills are composed of ciliated filaments which, in the more advanced species, are joined together either by ciliated or by vascular connections, to form folded and fenestrated plates (lamellae). The details of the arrangements are too complex and diversified to be described here, but in principle the beat of the cilia, which are arranged in a complex pattern on the sides and outer surfaces of the filaments, draws water towards and between the filaments, traps the food particles, and sweeps them over the surfaces of the ctenidia. The cilia are arranged in tracts in such a way that larger particles are swept to the outside of the mantle cavity and rejected, while the finer ones, which are more likely to be suitable for food, are conveyed to the mouth.

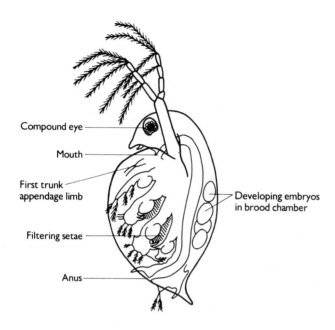

Fig. 4.7 A cladoceran.

This is the feeding method of the mussel (*Mytilus*) and oyster (*Ostrea*), which live attached to the substratum, where they filter the richly nutrient water of the shore and shallow sea (cf. Fig. 4.1). It is an unfortunate byproduct of human predation that in so doing they may collect polluted material (from sewage, for example) which, although harmless to themselves, may be infective to man if the molluscs are eaten before being cleansed by immersion for a few days in pure water. No less regrettably, they may themselves become victims of harmful chemicals added by us to their environment.

In this type of feeding the gills can be used with remarkable efficiency and flexibility. Regulation of the ciliary beat, in conjunction with adjustments of the gill lamellae, enables mussels to adjust the degree of particle retention in accordance with conditions in the water, without necessarily affecting the ventilation current and hence the respiratory exchange. More typically, however, bivalves live buried in the sand and mud of the shore, their microphagy ranging from suspension feeding to deposit feeding, in which particulate nutrient material on the sea bottom is ingested. This material may be

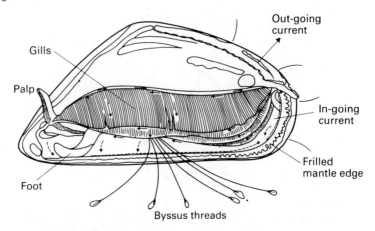

Fig. 4.8 View of the mantle cavity of the common mussel, *Mytilus edulis*, after removal of the left shell valve and mantle fold. Water is drawn in at the hind end below the gills, is strained through these and passes out above them, also at the hind end. Arrows on the gills show the direction of passage of food particles towards the palps which guard the mouth. Broken arrows indicate passage of excess particles back to the position of entrance where they are from time to time extruded by contractions of the adductor muscles. (After Orton (1912). *J. mar. biol. Assoc. U.K.*, **9;** from Yonge (1949). *The Sea Shore.* Collins, (London.)

living, or it may be dead, derived, as we have seen, from the waste products and dead bodies of other organisms.

Two examples of mud-living bivalves illustrate the close connection between suspension and deposit feeding (Fig. 4.9). *Mya arenaria*, living in the mud of estuaries, at depths of 30 cm or more, is a suspension feeder, reaching to the surface of the mud with two tubular processes (siphons). Water with suspended particles is drawn through one (inhalent siphon) and the outgoing filtered stream expelled through the other (exhalent siphon). *Scrobicularia plana*, living at depths of 15–20 cm on muddy shores, is a deposit feeder, with its two siphons separate, unlike those of *Mya*, and capable of greater elongation. The inhalent one draws in the surface deposits by ciliary action.

Worms, belonging to the class Polychaeta of the phylum Annelida, are also common in marine sand and mud. Some are macrophagous, ingesting small organisms with the aid of jaws; others, living in tubes, are suspension feeders, drawing in a stream of water and filtering it by some system of ciliated processes or tentacles. *Sabella*, with an anterior crown of stiff tentacles, is an example of this (Fig. 4.10). Like bivalves, it uses ciliary tracts to sort the particles, but economizes in energy by using some of the rejected ones for the

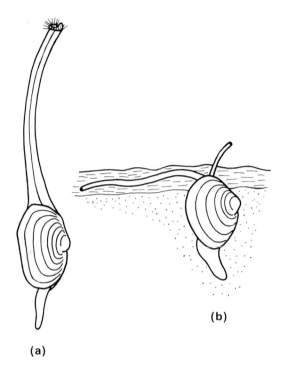

(b)

(a)

Fig. 4.9 (a) *Mya arenaria*, with siphons and foot fully extended. (b) *Scrobicularia plana*, showing use of siphons in feeding.

construction of the muddy tube in which it lives. More difficult to classify are those worms that swallow substratum material in bulk, extracting its organic content as it passes down the alimentary canal. Such an animal is *Arenicola*, the lugworm, which lives in a U-shaped tube (Fig. 4.11), drawing in sand from one end of it and ejecting it as the worm casts which are familar on sandy shores, lying close to the shallow pits which mark the incurrent end of the burrow. Is the lugworm microphagous or macrophagous? The question is not easy to answer, for the feeding mechanism of even so common an animal is not fully understood. Probably it is a selective feeder, using only the richly nutrient layer of sand and detritus at the head of its burrow.

The efficiency of filter feeding, expressed in such terms as the size of the particles retained, and the output of energy needed to secure them, raises issues which are of great intrinsic biological interest, and which also bear on the commercial rearing of invertebrates. The

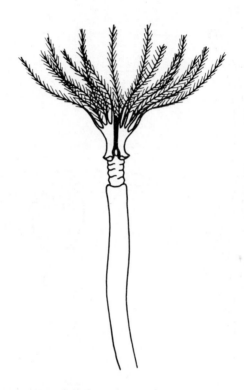

Fig. 4.10 *Sabella*, anterior end.

limited evidence available indicates that the larger copepods can retain particles of 5–10 μm, cladocerans 1 μm, and bivalves 1 μm to a few μm. This last range would serve to secure bacteria and small phytoplankton, but *Mytilus* is said to improve on this by filtering virus particles of 25 nm diameter, although the possibility of these aggregating to form larger particles is difficult to exclude. Quite exceptional efficiency is shown by the Appendicularia (Larvacea), a highly specialized group related to early forerunners of the vertebrates. Appendicularians (Fig. 4.12), making use of a secreted structure (the 'house'), with a complex filtering meshwork, can retain particles of 0.1 μm. These are then transferred to young herring and to plaice larvae, which feed extensively in the North Sea on the appendicularian *Oikopleura*. Marine biologists have also benefited from this efficiency, for the contents of the appendicularian 'house' have provided invaluable material for the study of the nanoplankton.

Data relevant to energy output by microphagous animals show that some filter feeders, including certain bivalves, which clear 15

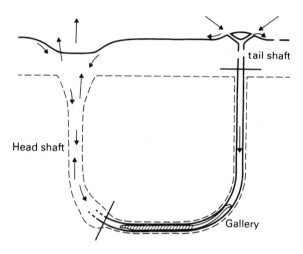

Fig. 4.11 Generalized diagram of a lugworm burrow, with the worm lying quietly. It moves to the head shaft to feed, and to the tail shaft to extrude faeces. Food is drawn down the head shaft, and the respiratory current of water passes out along it. (From Wells (1950). *Symp. Soc. exp. Biol.*, **4**, 127–42.)

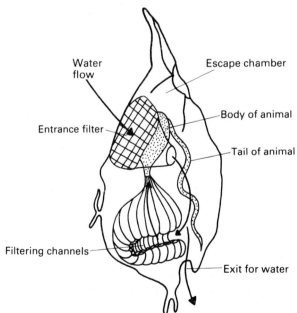

Fig. 4.12 *Oikopleura*, an appendicularian, in its 'house'. The animal leaves by the escape chamber when the filter becomes clogged.

litres of water for each ml of oxygen consumed, can cover their maintenance requirements (as estimated from their rate of oxygen uptake) if the water contains an organic equivalent of $0.13 \, \text{mg} \, l^{-1}$ of dried phytoplankton. This requirement is adequately met in the waters in which these animals normally live. However, some animals filter much smaller quantities of water, and it is open to question whether these, and particularly those clearing 1 litre or less, can meet their requirements solely by filter feeding. This, however, is less of a problem than might appear, for some filter feeders, and notably copepods, supplement this method by direct predation upon other members of the plankton.

Despite the suitability of the sea for life, plants and animals enjoying it must always have been subjected to the pressures of competition, which, in any habitat, lead them to extend their range and to exploit new sources of food. It is because of this that life has extended into estuaries, rivers and lakes, and from there, and also from the upper sea shore, on to dry land. We shall consider later some of the consequences of this extension, as they are expressed in plant and animal organization. As far as nutrition is concerned, the principles of food chain, trophic level, and pyramid of numbers remain unaltered (cf. Fig. 4.1). Moreover, in fresh water the same groups of organisms tend to be involved, including bacteria, diatoms, copepods, cladocerans and fish, although attack from terrestrially based predators is a more significant factor. In this connection, a good example of parallelism of adaptation is provided by certain birds. The small flamingo (*Phoeniconaias minor*), uses comb-like filters on its beak to strain out freshwater phytoplankton. A device similar in principle is used by the antarctic Prions (*Pachyptila* spp.), a group of small petrels which feed in dense flocks upon the smaller zooplankton in the surface water by straining them through lamellae fringing the bill. The flexibility of such feeding adaptations in response to selection pressure is exemplified by the larger flamingo, *Phoeniconaias antiquorum*. This uses its filter to extract small invertebrates from the mud of the same lake as that from which its smaller relative is drawing phytoplankton. Direct competition is thus avoided.

Some Terrestrial Relationships

The establishment of life on land involved the progressive development of adaptations in plants and animals which were integrated through their common involvement in producer/consumer relationships. The process was well in train nearly 400 million years ago, during the Devonian period (Table 4.1), when there was elevation of land and increasing aridity, with swampy conditions around the

Table 4.1 Table of geological periods.

Era	Period		Time since beginning of period in millions of years
Quaternary	Recent		(10 000 years)
	Pleistocene		2.5
	Pliocene		7
Tertiary	Miocene		26
	Oligocene		38
	Eocene		54
	Palaeocene		65
Mesozoic	Cretaceous		136
	Jurassic		190
	Triassic		225
Palaeozoic	Permian		280
	Carboniferous		
	Upper:	Pennsylvanian	325
	Lower:	Mississippian	345
	Devonian		395
	Silurian		430
	Ordovician		500
	Cambrian		570
Precambrian			

diminishing waters. The bryophytes, the simplest land plants (p. 163), which first appear in the Upper Devonian, have long been overshadowed by the vascular plants, but, if we can judge by their role today, they would from the beginning have served an essential ecological function in colonizing bare surfaces and contributing to a fertile soil. Ferns appear in the Devonian, and, with some 10 000 living species, still remain an important component of the terrestrial flora, albeit adapted especially to humid habitats. Lycopods (club mosses) appear in the Lower Devonian, and in the Carboniferous included aborescent forms, 30 m high, with secondary wood, which made a major contribution to the forests of the late Palaeozoic. Seed plants were also early on the scene. Pteridosperms (the now extinct seed ferns) appear in the lower Carboniferous, and made a major contribution to our coal deposits, while conifers (today the most widespread of the gymnosperms) also become well established during the Carboniferous. Angiosperms (flowering plants) appeared much later, the first authentic records being from the Cretaceous, but they proved to have important advantages, not least in their mode of reproduction, as we shall see, and so they quickly came to be the dominant members of the terrestrial flora.

The rapid development of this rich land flora, even though at first of primitive character, would have provided both food and shelter to promote the emergence of terrestrial animals. This is well shown in the establishment during the Devonian of terrestrial arthropods (spiders, mites, scorpions and millipedes); amphibians were evolving at the same time from fishes, doubtless taking a mixed diet from the Carboniferous swamps and from the edges of the water. Later, however, amphibians increasingly tended to return to the water, influenced, no doubt, by the evolution of reptiles, which first appear in the Permian. By this time the insects were becoming well established, primitive forms of these having appeared in the late Carboniferous. Birds and mammals appear in the Jurassic, and it is possible that flowering plants began to appear during this period, although this is uncertain. It is thought likely that the earliest angiosperms were woody plants, yielding forests which must have influenced the history of mammals, and may have been a factor in the evolution of bird flight. The history of grazing mammals was integrated with the somewhat later evolution of herbaceous plants, and particularly with the development of extensive grasslands during the Pliocene. Another aspect of these later phases is the integration of the habits of certain groups of insects with the reproductive methods of flowering plants; we shall return to this later.

One consequence of the emergence of terrestrial life has been the disappearing of filter feeding amongst terrestrial animals, although there is some analogy to it in the exploitation by spiders and birds of the rich insect fauna of the air, which can be regarded as a form of aerial plankton. The prime requirement for terrestrial animals has been the exploitation of bulk food, both plant and animal, by some form of macrophagy, with a consequent need for increased versatility of movement. The adaptations involved in macrophagy are too diverse to permit of many useful generalizations here. Vertebrates have depended upon their jaws and teeth, employing a variety of devices that culminate in the toothless beaks of birds, cunningly adapted for a diversity of food material and economically used also for intraspecific recognition, and in the complex and multipurpose dentitions of mammals.

Insects have been no less successful, owing much to the versatility of their mouth parts. Some use these as piercing organs for ingesting plant fluids, and are thus able to exploit plants while evading the considerable problem of the digestion of cellulose. Bugs (Hemiptera) have their mandibles and maxillae modified into piercing and sucking stylets. Those of aphids, forced between the plant cells, and strengthened by a sheath secreted around them, can penetrate the phloem and extract the sap with a precision that we have seen to have been exploited by human researchers (Fig. 4.13). So effective is this extraction that bugs often excrete a fluid sufficiently rich in

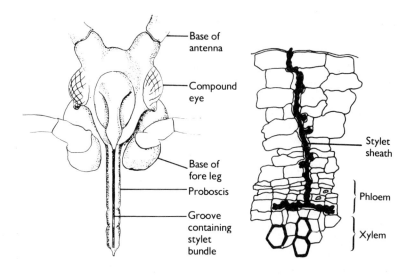

Fig. 4.13 (a) Anterior view of head and proboscis of an aphid, to show the groove in the proboscis in which lies the stylet bundle. (**b**) Stylet sheath laid down in the tissues of a plant by an aphid feeding on the phloem sieve tubes. (From Dixon, 1973.)

carbohydrate to be exploited by other animals. The excretion of a coccid bug living on tamarisk is thought to have provided manna for the Israelites, while the honey-dew excreted by aphids contributes to the nutrition of the ants which carefully foster them. Another fluid source used by insects is the nectar of flowers, converted to honey within the alimentary tract of bees, and then further exploited by man, especially in western countries prior to the discovery of the sugar cane.

There are carnivorous variants of these suctorial feeding methods, including the blood sucking practised by hemipterans, fleas, gnats and mosquitoes, all dependent upon the versatility of insectan mouth parts. The injection of an anticoagulant into the wound is a common accompaniment of this method of feeding, exploited by the causative parasites of malaria and sleeping sickness. The arachnids (including mites, ticks and spiders) use a different type of fluid feeding. They squeeze juices out of their prey with their crushing appendages, and secrete a saliva which initiates digestion within the prey's body.

Digestion

Ingestion (the intake of food) is only one aspect of animal nutrition,

for after the ingested material has passed into the alimentary tract it
is still part of the outside world. It must now be broken down into
molecules small enough to be absorbed into the body and, in due
course, to be passed along the metabolic pathways of respiration, or
to be built up into the structural elements and energy reserves which
constitute production. The breakdown is a hydrolytic process,
effected through the mediation of digestive enzymes. These are
sometimes secreted externally, to facilitate ingestion of food (as we
have seen, for example, in arachnids), but typically they function
within the alimentary tract, the structure of which is closely adapted
to their properties. These include maximal activity at characteristic
pH values, and a high specificity with respect to the substrate that
they attack, so that they can be classified into proteases (peptidases),
which hydrolyze proteins; carbohydrases which hydrolyze
carbohydrates; and lipases which hydrolyze fats (lipids).

Within these broad categories, further subdivisions can be made.
Proteases are distinguished into endopeptidases, which act upon the
more centrally situated bonds of peptide chains, and exopeptidases,
which attack only terminal bonds. Both categories are highly specific
with respect to the bonds that they can attack, a property exploited
in analytical studies of peptide structure. Thus endopeptidases
include pepsin and chymotrypsin, which attack bonds associated with
aromatic amino acids, and trypsin, which attacks bonds associated
with arginine and lysine residues. These enzymes are released in
vertebrates in inactive forms which are activated within the
alimentary lumen. The exopeptidases, which complete protein
breakdown, include carboxypeptidases, which remove terminal amino
acids with free carboxyl groups; aminopeptidases, which remove
those with free amino groups; and dipeptidases, which cleave the
single bond remaining in dipeptides (cf. p. 6).

Carbohydrases comprise polysaccharases, the commonest of which
are the amylases which hydrolyze starch and glycogen into the
disaccharide maltose, and the glucosidases, which hydrolyze maltose
and other disaccharides (e.g. sucrose and lactose) into glucose
molecules, and act also upon other simple carbohydrates, such as the
trisaccharide raffinose.

Lipases do not lend themselves to this type of subdivision. Many
are able to hydrolyze a variety of organic esters, although the fats
upon which they act are usually triglycerides, in which three
molecules of fatty acid are linked to one molecule of glycerol. The
breakdown of triglycerides is a stepwise hydrolysis, in which the ester
linkages are attacked successively, but, in contrast to the stepwise
breakdown of proteins and carbohydrates, the lack of specificity
permits one enzyme to complete the whole process.

The digestion of food by animals thus depends upon a sequence of chemical events, which must be linked with the passage of food through the alimentary tract, and with the absorption of the products. Not surprisingly, therefore, alimentary systems are often highly complex, but it must suffice here to examine a few basic principles of their organization.

First, it is obvious that digestion must have been intracellular prior to the appearance of metazoans, except in so far as enzymes were secreted externally, as we have seen them to be in bacteria and fungi. Intracellular digestion is characteristic of protozoans, which take up dissolved material by pinocytosis, or large particles into food vacuoles by phagocytosis. There is no important difference in principle between the two methods, both of which result in the enclosure of food material within intracellular vacuoles. Enzymes are then passed into these from lysosomes, probably often by the fusion of these with the vacuoles to form secondary food vacuoles. Soluble digestion products diffuse into the cytoplasm and undigestible residues are discharged through the cell surface, or sometimes, in more specialized protozoans, through a permanent cell anus (cytoproct).

The evolution of metazoan structure, and the consequent development of a digestive cavity, made possible the secretion of enzymes into this cavity, and thus the effecting of extracellular digestion. However, intracellular digestion still persists, not only in more primitive groups (notably in the coelenterates), but also to a varying extent in more specialized ones, and quite possibly throughout the animal kingdom. The carnivorous coelenterate hydra (Fig. 10.9) breaks down its prey (the water flea, *Daphnia*, for example) into small particles within four hours, by digestive enzymes secreted into the coelenteron from several types of secretory cell. The material is taken up into vacuoles within absorptive cells, where protein is further broken down and, it is thought, built up also into reserve protein. There is, however, little digestion of carbohydrates and fats; an illustration of the adaptation of enzyme equipment to the normal diet.

The persistence of intracellular digestion in more advanced animals is often associated with microphagy, no doubt because of the readiness with which the small food particles can be ingested by the digestive epithelium. This presumably has the advantage of economy in enzyme production, since the secretions can be brought to bear directly on the food, instead of being disseminated throughout the digestive tract. Moreover, it aids the exploitation of particulate food material, because the lumen is left free for a more continuous entry of food than might otherwise be possible. Bivalve molluscs are outstanding examples of this mode of digestion.

In arthropods, by contrast, digestion is almost exclusively

extracellular, although it is said that the final digestion of proteins is intracellular in some forms, notably the arachnids. In these respects the arthropods resemble the vertebrates, where, too, digestion is largely extracellular. It is not, however, wholly so. Studies of *in vitro* preparations of mammalian intestine show that dipeptides are taken up into epithelial cells in which they are finally broken down by intracellular dipeptidases. A possible explanation for this persistence of a primitive mechanism of digestion is that the time required for the complete breakdown of complex protein molecules is greater than can be provided during the passage of food along the intestinal lumen.

Digestive enzymes have afforded much scope for adaptation to particular types of diet, as we have seen in hydra, where a carnivorous diet is associated with minimal capacity for digesting carbohydrates and fats. In this type of adaptation, often of critical importance in ecological relationships, flexibility in diet is sacrificed to the efficient and economical utilization of the normal food. Insects provide many well-studied examples of this.

The omnivorous cockroach secretes a full range of enzymes, whereas the water-beetle *Dytiscus*, a specialized carnivore, secretes mainly proteases. The tse-tse fly, *Glossina*, feeding on blood, also relies mainly on proteases, whereas the closely related horsefly *Tabanus* has carbohydrases as well; a difference related to the feeding of the latter on decaying organic matter and plant juices as well as on blood. Similar differences may emerge even during the life cycle of a single species. Thus a butterfly larva, feeding on plant cells, may have a full range of enzymes, whereas its adult, feeding only on nectar, secretes only a few carbohydrases. Such adaptations make for economy of energy and for the avoidance of competition for food resources within a community, both of these being factors which will be favoured by strong selection pressure.

Adaptation of a different type is found in the clothes-moth larva, *Tineola*, which is one of the few insects able (to our annoyance) to digest the very resistant protein keratin (the bird-lice, Mallophaga, can also do so). The protease of the larva seems no different from that of other insects. The adaptation here is a specialization of the alimentary tract, one region of which maintains a low oxidation-reduction potential. This results in the strong S–S bonds, which bind the folds of the polypeptide chains, being opened by the formation of SH groups; the protein can then be readily digested.

Digestion of cellulose

One capacity which has attracted much attention because of its importance in making full use of primary production, is the digestion of cellulose. Bivalve molluscs, exploiting ciliary and deposit feeding, have a wide range of carbohydrases, including a laminarinase, which is probably a complex of glycosidases. Cellulases have also been found in them, *Mytilus* and other filter feeders having higher concentrations than deposit feeders; perhaps, it has been suggested, because of the higher proportion of cell walls ingested by filter feeders. Amongst other animals which have been thought to digest cellulose are the snail *Helix*, which has been credited with an extraordinary range of enzymes, as many as thirty having been listed. Other examples, all with specialized feeding habits, are the bivalve *Xylophaga*, which bores into floating timber; *Teredo*, the highly modified bivalve ship-worm; the wood-boring isopod crustacean, *Limnoria* (the gribble); and certain other terrestrial isopods.

There is, however, a major difficulty in evaluating these conclusions. Cellulolytic bacteria are widely distributed, so that the cellulase activity could be due to a microflora living within the alimentary tract, a possibility not excluded by many earlier records. Bacteria are now thought to be implicated in *Helix* and in terrestrial isopods, and to contribute to the very high cellulase activity found in the gut contents of *Mytilus californianus*. It seems likely that only a very few animals, at most, have succeeded in evolving a cellulase system of their own, perhaps because the ubiquity of cellulolytic micro-organisms has satisfied this ecologically important requirement. However this may be, there are certain groups of animals which have developed such dependence upon micro-organisms to a very high level of specialization, their consequent ability to utilize cellulose making a significant contribution to energy flow. This could have begun with the chance association of ubiquitous micro-organisms with a host species that had access to cellulose-rich diet, leading to the establishment of interspecific symbiotic relationships of remarkable complexity.

Examples of this are provided by termites; insects with feeding habits that give them great economic importance. Some of them utilize the broken-down plant material of humus, thereby filling an ecological niche that is occupied elsewhere by earthworms. Some, however, actively ingest and directly utilize timber and other plant products. Wood may contain up to 38 % lignin, which is probably never digested by animals, but a large amount of it is cellulose. This requires for its digestion cellulases and β-glucosidases (analogous to the amylase and maltase required for the digestion of starch). Some wood-eating termites, unable to secrete these enzymes themselves, make use of flagellate protozoans which inhabit their gut (Fig. 4.14) in quantities so vast that they may account for as much as one-third of the weight of the host insect.

Some of these flagellates (Fig. 4.15) feed upon bacteria which are

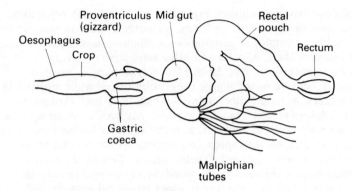

Fig. 4.14 Alimentary canal of a *Zootermopsis* worker, a damp-wood termite; cellulose-digesting protozoans inhabit the rectal pouch.

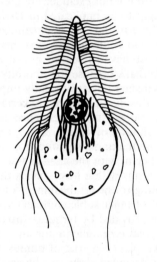

Fig. 4.15 *Trichonympha collaris*, a protozoan from the alimentary canal of *Zootermopsis*. (Length about 200 μm.)

also present in the gut, or they may absorb the surrounding nutrient solution, but others can be seen to ingest fragments of wood. These they break down, discharging the digestion products which are then taken up by the insect. The protozoans and their termite hosts are completely dependent upon each other. The termites cannot live without their associated flagellates, while these are found only in the termite gut, where they function as obligate anaerobes, fermenting the cellulose to organic acids which are oxidized by the host after it has absorbed them.

Another solution to the problem of cellulose digestion is found amongst the fungus-growing termites, as exemplified by *Macrotermes natalensis*. This maintains and ingests cultures of a fungus (*Termitomyces*) which it rears in its nest upon chewed but undigested plant fragments. The fungus, which is thought to occur only in termite nests, produces a cellulase (C_1-enzyme) which breaks down crystalline cellulose. Its action is followed by a C_2-enzyme, which is active against non-crystalline cellulose and some of its degradation products, and then by β-glucosidases. The termite is completely dependent upon its fungus as a source of the essential C_1-enzyme, but is able to produce the other enzymes itself. It has been suggested that such extension of food resources, by exploiting external supplies of an essential enzyme, and originating by chance associations as just envisaged, may be widespread amongst animals that feed on litter, detritus or dead wood.

A relationship similar to the use of intestinal flagellates by termites, but one that is much more important in the utilization of primary production, is provided by the herbivorous mammals of the order Artiodactyla, which comprises the even-toed ungulates, such as cattle, sheep, camels, giraffes and deer. All of the order, with the exception of pigs, peccaries and hippopotamuses, swallow their food into a highly modified and multi-chambered stomach (Fig. 4.16); from this it is regurgitated, chewed at leisure, and then passed back

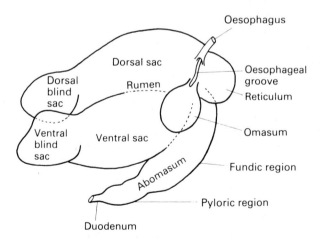

Fig. 4.16 The multi-chambered stomach of an adult ruminant. In the young animal the milk bypasses the reticulum and omasum and passes directly into the abomasum through the oesophageal groove, which closes to form a channel during suckling.

into the stomach. This consists of the rumen, reticulum (these two regions provided the original wrapping for genuine Scottish haggis), omasum and abomasum, the gastric juice being secreted in the last of these.

The food, rich in cellulose, is largely broken down in the rumen, which is a fermentation chamber of 100 litres average capacity in cattle, and 10 litres in sheep. It is partly filled by a fluid in which live symbiotic bacteria and ciliate protozoans in vast numbers, averaging 10^{10} bacteria and 10^5 protozoans per cm^3. Mammals do not secrete cellulolytic enzymes; here, as in termites, it is the microflora and microfauna, obligate anaerobes in a largely anoxic environment, which ferment the cellulose into carbon dioxide and fatty acids. Most of the latter are absorbed by the host's stomach, but some are used by the micro-organisms themselves. Excess gases (carbon dioxide, and methane formed by its reduction) are removed by belching.

Relationships within the rumen are complex, and constitute a miniature ecosystem. Thirty to fifty species of ciliate have been identified in it, some feeding on starch and sugars, while some (e.g. *Diplodinium*) ingest cellulose and break it down. The role of the bacteria is not entirely clear, although it has been thought that the cellulolytic capacity of *Diplodinium* might be due to bacteria living within it. The similarity of the ruminant system to that of termites is sufficiently obvious, and it is an instructive example of parallel evolution, in which unrelated groups of animals evolve similar adaptations in response to similar environmental demands. There is, however, one important difference. Termites do not usually digest their micro-organisms, but artiodactyls do, and are estimated to obtain up to 20% of their protein nitrogen from them, as well as a considerable supply of B vitamins.

Other mammals make use of cellulose through the action of micro-organisms living in their enlarged caecum and colon. Examples are the Perissodactyla (odd-toed ungulates, such as the horse, in which the large intestine and rectum constitute up to one-half of the total volume of the alimentary tract), rodents (rats and mice), and lagomorphs (rabbits and hares). The system functions in principle like that of artiodactyls, but less efficiently, in part because the range of species of micro-organism is smaller, and in part because these are too far back in the alimentary tract to be digested as an additional source of food. However, lagomorphs partly overcome this drawback by eating their faeces (coprophagy), the contents of the caecum and colon being thus recycled through the alimentary canal.

5 Respiratory systems

Gaseous Exchange

Animal consumers, like plant producers, depend for the release of energy within their bodies upon the complex of processes that we have referred to as cell respiration, but there is an important difference between the two in this respect. Because animals usually move about, and because they have complex organ systems that have no parallel in plants, they usually have structures, of varying complexity, which are responsible for external gaseous exchange. These structures constitute the respiratory system, and the functions that they fulfil are sometimes referred to as external respiration. Further, oxygen and carbon dioxide are transported within the body, to and from the respiratory system, by some form of circulatory system, in all but the smallest and simplest of animals. Both systems show a diversity of adaptations to the demands imposed by the environment.

Perhaps the most important single factor influencing those adaptations has been the movement of animal life from the seas into freshwater habitats and onto the land. This has posed many respiratory problems, resulting from differences in the gaseous composition and physical characteristics of air and water, and, to a less extent, of sea water and fresh water. In one sense, all animal respiration is aquatic respiration, for the gases concerned pass between the external medium and the body by diffusion through an aqueous layer, even when the respiratory organs are enclosed within the body. This is as true of the mammalian lung as it is of the gills of fish. Differences in the properties of air and water, however, have important effects on the ease with which the gaseous exchanges can be effected.

Air contains some twenty times as much oxygen as does water saturated with air; aquatic animals must therefore pass correspondingly more of the medium over their respiratory surfaces. Their difficulties are further increased by the high density and viscosity of water (respectively about 1000 times and 100 times that of air), and by the diffusion rate of oxygen being very many times lower in water. Other difficulties arising in aquatic life are that the solubility of oxygen in water is reduced by a rise in temperature (from 9 ml l^{-1} at 5°C to 5 ml l^{-1} at 35°C) and by a rise in salinity. Finally (and this is particularly, although not exclusively, a problem of freshwater life), local conditions may set up zones poor in oxygen, as, for example, in swamps during dry season, or at the bottom of muddy ponds, where poor circulation of the water can impede the replacement of the oxygen that is used up by detritus-living organisms. It is therefore not surprising to find that the activity of the respiratory organs of fishes accounts for about 20–30% of their resting oxygen consumption, whereas in mammals the corresponding figures are only 1–2%. Here, perhaps, is one of the advantages gained by terrestrial life. It is generally held that a major factor in the movement of animals onto land, at least as far as vertebrates are concerned, was the stress of oxygen deprivation in tropical swampy conditions, with a consequently high selection pressure favouring the exploitation of aerial respiration.

We have said that the gaseous exchanges of external respiration take place by diffusion, but passive diffusion is only adequate in the very smallest animals or in those, such as sponges and coelenterates, in which the external water bathes thin epithelial layers. With increase in size, diffusion becomes progressively less effective, and for two reasons. One is that with increase in size of the body, the ratio of surface area to volume declines, a fact of geometry which profoundly influences many aspects of animal organization, and, indeed, of plants as well. In the present instance, increase of size progressively decreases the availability of oxygen for the internal tissues, in the absence of compensating adaptive devices. The other reason is simply that the slow rate of diffusion of oxygen through the body fluids will be inadequate to support metabolism.

Adaptive modification of the body shape can to some extent circumvent these difficulties. A sphere provides the smallest surface/volume ratio for any one size of body. Flattening improves the ratio, which is why flatworms (platyhelminths) can rely upon gaseous exchange through their general body surface, in conjunction with the maintenance of a relatively low metabolic rate. But increased size and activity, and the diversification of modes of life which these make possible, have usually required the evolution of respiratory organs in the form of structures called gills, lungs or tracheae. As with other aspects of animal organization, these terms are sometimes rather loosely used, and it is important to remember that superficially similar respiratory adaptations have been evolved quite independently in unrelated groups. As always, diversity in biological organization is constrained by the limited range of solutions available for a particular problem.

Respiratory Organs

The primary function of respiratory organs is to increase the respiratory surface without interfering with the other activities of the animal, and without defeating their purpose by unduly increasing its volume. Usually, however, they help to meet another requirement. Inward diffusion of oxygen is aided by maintaining the maximum possible gradient of oxygen tension across the respiratory surfaces. Internally, this is commonly secured by the circulatory system, which removes oxygen as fast as it enters, and which may sometimes (in earthworms, for example) permit adequate respiratory exchange without the use of specialized respiratory organs. Externally, the gradient is promoted by maintaining movement over the respiratory surface; this is achieved sometimes by the movement of the animal itself through the water, but more usually by ventilation movements, which displace the water or (in terrestrial animals) the air.

Gills, in principle, are vascularized extensions of the body surface, sometimes exposed, but often, because of their delicate and vulnerable folding, protected and concealed. They are characteristic of aquatic animals, partly because there is no problem in keeping them moist, and partly because the high density of water supports their folds and maintains the efficiency of their respiratory area. On land, the gills could not be supported by air, nor could they easily be kept moist; they would collapse, and their respiratory efficiency be lost, as is obvious in fish. Eels can only move over land for considerable distances because the opening into the respiratory (opercular) chamber is greatly reduced, and the gills kept moist and functional.

Versatility such as this is found also in invertebrates, permitting primarily aquatic groups to extend their lives onto land. Examples are seen in malacostracan crustaceans. These typically respire by gills borne as extensions of some of their limbs, either as plates or as filamentous structures. In the decapods (e.g. crayfish, crabs) the filamentous gills are protected within a gill chamber (branchial cavity) (Fig. 5.1), formed by an extension of the body surface, ventilation being maintained by movements of the carapace and by a bailing action of the scaphognathite, a paddle-like structure borne on the second maxilla. Some crabs are able to live on land for several days at a time by retaining water within this cavity, but they have to replenish it at intervals so that they are amphibious rather than truly terrestrial. A further step is to replace the water in the chamber by air, as is done in the land crab *Birgus*, respiratory diffusion taking place through the lining of the branchial cavity, which is well vascularized. Essentially, the cavity has become a lung,

96

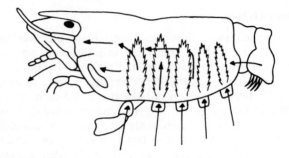

Fig. 5.1 Path of water circulation through gill chamber of crayfish. Water enters at bases of legs and at posterior margin of carapace.

the term given enclosed respiratory organs for breathing air. A similar transformation has taken place in the evolution of the pulmonate gastropods, as exemplified in the garden snail (Fig. 5.2), where the molluscan mantle cavity, which typically would contain gills, has here been converted into a lung (pulmonary cavity), with loss of the gills and with exchanges taking place through the vascularized lining. Air enters through a narrow opening, the pneumostome, movements of the body providing for ventilation.

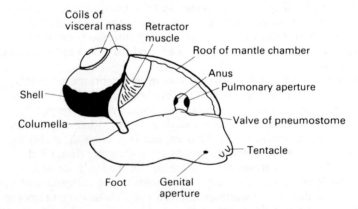

Fig. 5.2 Land snail (*Helix*), partially dissected to show pulmonary aperture (pneumostome).

The gills of fish are developed out of the alimentary tract as folds of the pharyngeal wall. Typically, water is drawn in at the mouth and passed over the gills through gill slits in the wall of the pharynx, although in the flattened bottom-dwelling rays it enters by the spiracle, which has moved dorsally, and leaves ventrally by the gill slits. Ventilation has been well analyzed in teleost fish, and shown to depend in principle upon the cyclic action of two pumps (buccal and opercular), separated by the gills, and functioning in conjunction with valves (Fig. 5.3). Movement of the jaws and of the floor of the mouth enlarges the buccal cavity, the consequent fall of pressure drawing water in by suction, with the buccal valve open. Outward movement of the operculum (gill cover), initiated slightly later in the cycle, reduces pressure in the opercular cavity below that in the buccal cavity so that water passes into the former across the gills. This flow is then increased by a rise in buccal pressure as the size of the buccal cavity is reduced. The cycle is completed by a rise in opercular pressure which opens the opercular valve and forces water out of the opercular cavity.

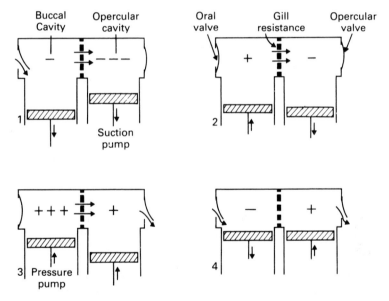

Fig. 5.3 Diagrams to show the double pumping mechanism for ventilating the gills of fishes. *1* and *3* are the major phases of the cycle; *2* and *4* are brief transition phases. (From Hughes and Shelton (1962). *Adv. comp. Physiol. Biochem.*, 1.)

Subtle features, including the phasing of the pump operations, the nature of the gill resistance, the movements of the valves, and adjustments of individual gill arches, ensure that flow of water across the gills is more or less continuous throughout the pumping cycle. Respiratory efficiency is further increased by the circulation of the water and blood being in opposite directions, establishing what is called in engineering practice a counter-current (multiplier) system (Fig. 5.4). Given that the rates of movement of the two currents are approximately the same, this ensures that where the two paths of flow are in contact, there will be a maximum diffusion gradient between them, yielding, under experimental conditions, an 80% extraction of oxygen. If both streams were moving in the same direction, the gradient would become progressively smaller along the region of contact; this arrangement has been shown to yield only 10% extraction.

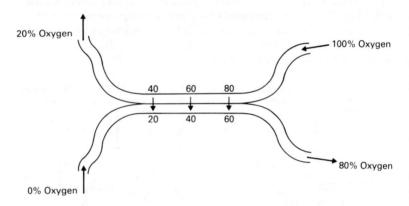

Fig. 5.4 Diagram to show the principle of oxygen extraction by counter-current flow. Bold arrows indicate the direction of flow; short arrows indicate direction of diffusion.

This respiratory system of fish lends itself to adaptive modification, related to the environment. For example, in bottom-living teleosts (e.g. plaice, sole) the mouth can move very little, and ventilation is mainly effected by slow suction exerted by the opercular pump. This, it is thought, may aid the camouflage of these protectively coloured animals as they rest on the sea bottom, and may also prevent sand from entering the ventilation stream. A contrasted example is the mackerel, an active and continuously swimming fish. Respiratory movements cease while it is swimming, ventilation being effected solely by water entering the mouth during forward movement. It is probable that the energy thereby saved is transferred to the swimming musculature, with a consequently enhanced efficiency of movement.

The prime respiratory requirement for the passage of fish onto land was the development of lungs, supplemented by internal nostrils and a buccal force pump. So much we can infer from the fossil osteolepid fish, from which land vertebrates are believed to have arisen, and from the living lungfish (Dipnoi), which, although not on the direct line of ascent, are closely related to it. The lungs of these animals develop as ventral outgrowths of the pharynx, air being gulped into the buccal cavity and driven into the lungs by contraction of the buccal cavity and pharynx. Warm water contains less dissolved oxygen than cold water, which is probably why certain tropical teleosts also gulp air, and have specializations of the pharyngeal epithelium for respiratory exchange. It is easy to see how true internal lungs, which develop in vertebrates as outgrowths of the anterior end of the alimentary tract, could have arisen from similar habits.

A buccal force pump is used also by amphibians, the lungs of which, however, are simple sacs, providing only a small surface for gaseous exchange. The system is supplemented by vascularized areas of the buccal lining and the skin, the proportions of the three components varying with habitat. Thus the newt (*Triton*), which spends much of its time in the water, and which uses its skin and its lungs, relies mainly upon the skin, which is estimated to contain some 75 % of its respiratory capillaries. The reverse obtains in the tree frog (*Hyla*), in which the lungs contain 75 % of them. Some fully aquatic salamanders lose their lungs altogether, relying mainly upon the skin capillaries; the consequent reduction of buoyancy probably aids their concealment. In general, we see in these and other amphibians a group imperfectly adapted to terrestrial life; respiratory limitations contribute to their sluggish and evasive habits, with their permeable skin making them especially prone to desiccation (see Chapter 7).

It is within the reptiles that the foundations of a successful terrestrial respiratory system are laid. The internal surface of the lungs is increased by folding, accompanied by penetration into the lungs of branches (bronchi) given off by the main trachea. This obviates respiratory dependence upon the skin. Air is now drawn in by movements of the ribs and body wall, acting as a suction-pump mechanism, while the elasticity of the lungs, always one of their important properties, provides for expiration, aided by movements of the body wall.

The highest degrees of respiratory efficiency are attained in birds and mammals, making in both groups a vital contribution to the homeothermy that we shall examine later. Air is drawn into the lungs by suction; in birds by expansion of the cavities of the thorax and abdomen, in mammals by expansion of the thorax, aided by movements of the muscular diaphragm which closes the thoracic cavity behind (below in man, of course). In birds (Fig. 5.5) the internal surface area of the lungs is relatively small, but this is

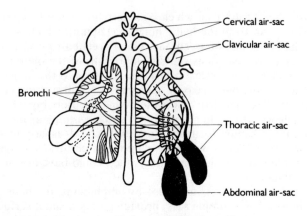

Fig. 5.5 Diagram of lungs and air-sacs of a pigeon, seen from the ventral side on the left, dorsal on right. On the left side only the ventral surface of the lungs and the expiratory bronchi and air-sacs are shown (dotted). On the right are the inspiratory bronchi and air-sacs (black). (After Brandes and Ihle, from Young, 1962.)

because the air passes not only into them, but also from the bronchi into a series of air sacs, structures which are peculiar to birds. Air passes first into posterior inspiratory sacs, and then into anterior expiratory ones, which may extend into the bones. It may then be expelled, or recirculated through the whole system. Exactly how this system works is not entirely clear, but it seems that much respiratory exchange takes place in the ducts (parabronchi) which connect the bronchi with the sacs. Clearly, however, the arrangement ensures a direct flow of air right through the respiratory system, presenting a marked contrast with that of mammals. Here, because there is no through flow, and because the lungs cannot be completely collapsed by their own elasticity or by the movements of the thorax, they always retain some residual air, part of which is in the small cavities (alveoli) in which the gaseous exchange takes place. In man, even after the most vigorous ventilation, there remains some 1.2 litres of residual air out of a total lung capacity of about 6 litres. It seems likely that ventilation yields a more efficient uptake of oxygen in birds than in mammals, a difference well correlated with the higher metabolic rate of the former, and with the high energy demand made by flight.

Respiratory Pigments

We have mentioned the low solubility of oxygen in sea water as presenting a problem in the respiration of marine animals, but it is, in fact, a problem that runs right through the animal kingdom, for if oxygen is to be transported round the body it can only be in a saline biological fluid (except in those animals, to be mentioned later, that have evolved tracheal systems). The active life of the larger and more advanced animals would be quite impossible if they had to rely upon the transport of dissolved oxygen, for the most that could be held in such solution is about 0.5 %. It is this which accounts for the evolution of the coloured substances called respiratory pigments, which have the property of combining loosely with oxygen at the respiratory surfaces and readily giving it up to the metabolizing tissues.

The best known of these is the red pigment haemoglobin, present throughout the vertebrates (with the exception of some antarctic fish), and in a number of invertebrates as well. As a result, mammalian blood may contain about 25 % oxygen by volume, birds 20 %, reptiles 10 %, amphibians 12 %, and fish 10 %, whereas vertebrate blood plasma, lacking the haemoglobin, contains only some 0.3 %, this being in physical solution.

Haemoglobin is a conjugated protein in which the protein, globin, is combined with an iron-porphyrin compound, haem (Fig. 5.6). The ferrous iron atom is bonded by four of its coordination valencies to nitrogen atoms of the porphyrin molecule; the fifth valency attaches the iron-porphyrin to the globin molecule, while the sixth carries either a water molecule (haemoglobin) or an oxygen molecule

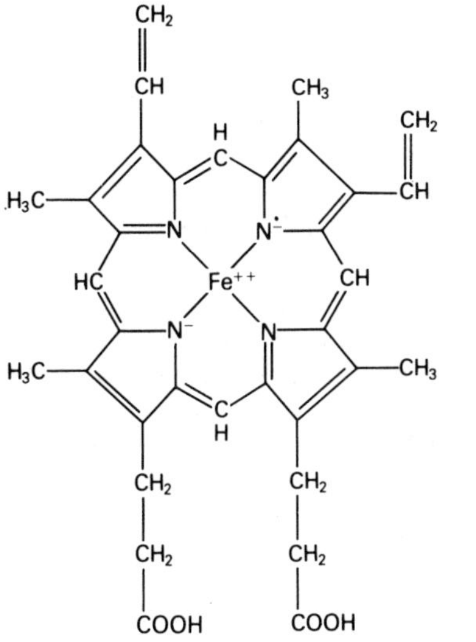

Fig. 5.6 Chemical structure of haem.

(oxyhaemoglobin, see below). The pigment is particularly characteristic of vertebrates, but is widely, although sporadically, distributed in various invertebrates, such as lugworms, earthworms, and the larvae of chironomid flies. The haemoglobins of most vertebrates are polymers, composed of four haemoglobin subunits, and with molecular weights of about 68 000, but the jawless cyclostomes (surviving members of a very ancient group which preceded the jawed fish) are exceptions, the lamprey having only one unit, with a molecular weight of about 17 000, and *Myxine* (the hagfish) having two.

All vertebrates, however, possess another monomeric globin, called myoglobin, which is present in their muscles. This pigment, like haemoglobin, combines reversibly with oxygen, and, having a high affinity for oxygen, can aid its transfer from the blood to the muscle cells. It is likely that myoglobin and the haemoglobin monomer are primitive features of the vertebrates, derived from a common ancestral molecule, and that the tetrameric form of haemoglobin arose during the early evolution of fish, perhaps because it gave increased efficiency in oxygen transport. The condition in invertebrates shows much variation. A dimeric form is found in certain polychaete worms, while the haemoglobin of lugworms and earthworms may have as many as 180 subunits, giving molecular weights of some 3 000 000.

It will be noticed that haemoglobin closely resembles chlorophyll and the cytochromes (p. 35, and Fig. 2.2), and this accounts for its wide distribution. The biochemical pathways needed for the biosynthesis of metallo-porphyrins seem to be universal, so that the need for oxygen carriers could often have been met by the association of already existing iron-porphyrin complexes with suitable proteins. This is another example of similar biochemical adaptations evolving independently in unrelated groups as a consequence of the limited range of possibilities available to them.

The haem moiety of haemoglobin is always the same, but the globin varies, permitting variation in the properties of the molecule, which are customarily demonstrated by plotting the percentage saturation of the haemoglobin after exposure of the blood to different oxygen tensions or partial pressures (pO_2). (It will be recalled that in a mixture of gases, the pressure exerted by each is determined by its percentage.) The resulting curve is called an oxygen dissociation curve. It is an expression of the ability of oxygen molecules to combine with one or more of the four iron atoms of the haemoglobin molecule, producing different degrees of saturation. The combination is a loose one, forming oxyhaemoglobin, which is not an oxidation product, so that oxygen is readily yielded up again from it.

Dissociation curves for a typical mammalian haemoglobin are shown in Fig. 5.7. Their shape is of great biological importance, for it reveals that there is a substantial unloading of oxygen over a relatively narrow range of oxygen tensions, which corresponds closely with the actual range in the tissues. Unloading in these is further aided by the effect of carbon dioxide tensions, increase in which shifts the curve to the right (Bohr effect); this results in unloading being greater in regions where metabolism is active.

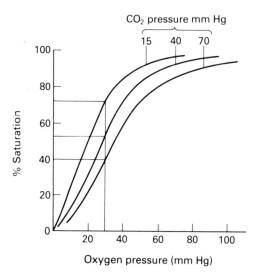

Fig. 5.7 Oxygen dissociation curves for mammalian blood at various CO_2 pressures illustrating the Bohr effect, the curve shifting to the right with increased pressure. Note that the percentage saturation varies from 40 to 72 as the CO_2 pressure varies from 70 to 15 mm Hg. (From Chapman, 1967.)

A convenient index of the oxygen-carrying capacity of haemoglobin is the oxygen tension at which the haemoglobin is half-saturated (i.e. 50% oxygenated), but it must be remembered that this value (p_{50}) is an approximation, for it will vary, not only with pCO_2, but also with temperature and with pH. Human blood, with a p_{50} of about 25 mm Hg, is said to have a low oxygen affinity. The same is true of the haemoglobin of salmon and trout, with values of about 20 mm Hg, but carp haemoglobin, with a value of about 5 mm, is said to have a high affinity, since it yields up its oxygen less readily. The difference between these fish is correlated with their habitat; salmon and trout require well-oxygenated water, whereas carp can live in water of low oxygen content, such as is found in stagnant ponds.

The effect of temperature is also of great adaptational importance. Thus the haemoglobin of the eel is far more readily saturated at 30°C than is that of the ray. The difference is correlated with the greater range of temperature encountered by the eel in fresh water, and its migration movements over land, in contrast to the much more stable marine environment of the ray.

The adaptive flexibility of haemoglobin, attributable, as we have seen, to its protein component, is evident also in invertebrates. In some of these it has a very high affinity, and it has been supposed that in such cases it acts as an oxygen store for use in conditions of oxygen deprivation, rather than simply for transport. This view, however, has not always stood up to closer investigation, *Arenicola* providing a case in point (Fig. 5.8). Its haemoglobin, with a p_{50} ranging from 2.0 to 8.3 mm Hg, has a high oxygen affinity, yet the amount of oxygen that it could hold as a store would only suffice for about 21 min of activity. The animal can draw oxygen into its burrow (Fig. 4.11) even when the tide is out, so that it does not have to depend on stored oxygen over long periods, although a small store might perhaps be of use to it. In fact, it seems more likely that the real value of the haemoglobin here is to take up oxygen from the sand around the burrow, which, during low tide, contains oxygen at a tension of about 6.7 mm Hg.

An instructive contrast to this is provided by another burrowing polychaete worm, *Nephtys*, which differs from *Arenicola* in being an

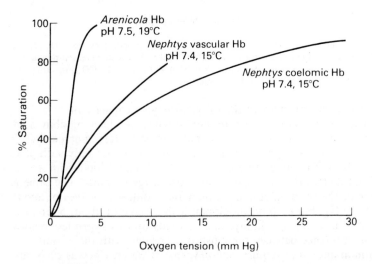

Fig. 5.8 Oxygen dissociation curves for haemoglobin of *Arenicola* and *Nephthys*. (From Jones (1955). *J. exp. Biol.*, 32.)

active animal that moves vigorously through the substratum. This probably accounts for its haemoglobin releasing its oxygen over a much wider range of oxygen tensions than does that of *Arenicola*, so that it is essentially a low-affinity pigment, although its p_{50} value is relatively low at 7.5 mm (Fig. 5.8). It is supposed that the worm has abundant oxygen while the tide is in, and that it has to reduce its metabolic rate when the tide is out, perhaps becoming partially anaerobic as the ambient oxygen tension rapidly falls.

One other example must suffice to illustrate the important ecological implications of the properties of haemoglobin, this time taken from the Pulmonata, the group of gastropod molluscs in which, as we have seen, the mantle cavity functions as a lung for aerial respiration. *Planorbis corneus* (Fig. 5.9a), a common freshwater pulmonate, is unusual in this group in having dissolved in its blood a haemoglobin with a high oxygen affinity. Like many other aquatic molluscs, it probably respires through its body surface, and it also has a secondarily developed gill, but it relies largely upon air which is taken into its lung when it is at the surface. It feeds by diving and browsing on bottom deposits, its haemoglobin continuing to take up oxygen from the lung as the oxygen tension in this falls, and thereby permitting a prolonged dive. Another freshwater pulmonate, *Lymnaea stagnalis* (Fig. 5.9b), does not possess haemoglobin, and it is probably because of this that its dives are shorter than those of *Planorbis*, and that it feeds upon vegetation near the surface of the water. Haemoglobin is thus not an essential requirement for diving pulmonates. Both species seem to be well adapted for an aquatic life, and it may be that the differences in their modes of respiration help them to avoid coming into direct competition with each other.

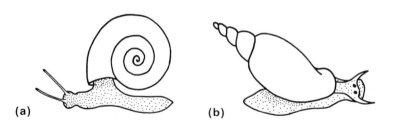

Fig. 5.9 (a) *Planorbis corneus.* (b) *Lymnaea stagnalis.*

In addition to haemoglobin, three other respiratory pigments are found in invertebrates. These will be referred to only briefly, since their functioning follows the same general lines as those just described. One is haemerythrin, found in cells (usually in the coelomic fluid) in one small marine phylum, the Sipuncula

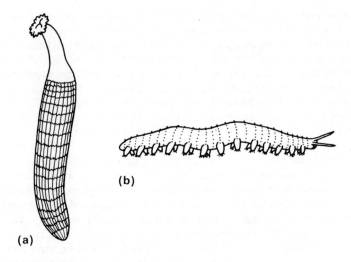

Fig. 5.10 (a) *Sipunculus nudus.* (b) *Peripatus.*

(Fig. 5.10a), and in a few other scattered species. Although red in colour, it differs from haemoglobin in being an iron-containing protein without any porphyrin. Molecular weights of around 66000 and 119000 have been reported. Chlorocruorin is a green pigment found in solution in some polychaete worms; it resembles haemoglobin in consisting of a protein conjugated with an iron-porphyrin haem, but has a consistently high molecular weight of around 3000000. The most important of the three pigments is haemocyanin, found in solution in a number of arthropods and molluscs. This pigment, which is colourless when deoxygenated and blue when oxygenated, differs from the other two, and also from haemoglobin, in being a protein with copper taking the place of iron. As with haemerythrin, there is no haem, the metal being attached directly to the protein. The molecular unit contains two copper atoms, and oxygen combines with both of these to form the oxygenated blue compound. The units combine to form complex molecules with very high molecular weights, ranging from several hundred thousand in arthropods to several million in molluscs. The dissociation curve shows a typical Bohr effect, although this may be reversed at greater acidities because of an increase in the stability of oxyhaemocyanin.

Tracheal Respiration

Arthropods containing haemocyanin include *Limulus* (Xiphosura), a few arachnids, and many crustaceans, but it is not found at all in insects. Indeed, these latter animals lack, in general, any respiratory pigment, although haemoglobin is found quite sporadically in a few; the larvae of chironomid gnats, which have it in solution in the blood, are a well-known example. Insects, extraordinarily successful and active animals, are able to function without an oxygen-transporting pigment because they have an entirely different mode of conducting oxygen through the body: the system of delicate tubules called tracheae. Tracheal respiration, although especially characteristic of insects (Fig. 5.11), has been evolved also by other arthropods. It is found in the archaic arthropod *Peripatus* (Fig. 5.10b), in myriapods (millipedes, centipedes and symphylans), and in many arachnids (including spiders, mites, ticks and harvestmen). There is no very close genetic relationship between these various groups, and it is certain that tracheae must have evolved independently on many occasions. We may infer that the mechanism is particularly appropriate to terrestrial animals with a tough and largely impermeable body surface (although this is not true of *Peripatus*). As we shall see, however, it does impose certain limitations in organization.

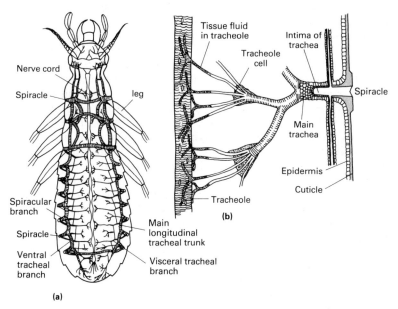

Fig. 5.11 (**a**) Generalized insect, showing arrangement of tracheae. (**b**) Details of arrangement simplified. The larger branches are supported by circular or spiral cuticular thickenings. The smaller ones develop within tracheole cells and are without support. (From Meglitsch, 1972.)

Tracheae (Fig. 5.11) are ectodermal ingrowths, lined with cuticle, and with external openings called spiracles (or stigmata), which are presumably the original sites of invagination. The danger of excessive water loss taking place through these has led to the development of specialized closing mechanisms which are controlled by many different means. In some instances (e.g. the cockroach) they open when carbon dioxide production is increased, thereby increasing respiratory exchange just when it is required.

The tracheae end blindly in very minute and delicate tracheoles with permeable walls that either surround the cells, or end within them. Here their functioning is aided by the permeability of the tracheoles, which not only permits gaseous exchange, but also results in water entering them from the surrounding tissues. Oxygen shortage at times of exceptional activity will result in metabolites accumulating in the tissue fluids; their osmotic pressure will rise, water will be osmotically removed from the tracheoles, and a greater space thereby created for the diffusion of oxygen to where it is most required.

The functioning of tracheal systems is governed by the properties of air and water which we have already discussed. The diffusion rate of oxygen in air is sufficiently high to provide a supply of oxygen to the tissues, provided that the animals satisfy certain conditions of size and activity, related to the geometrical constraints mentioned earlier. If linear dimensions increase by a factor of 10, the rate of diffusion of oxygen can also do so, but this must eventually fail to meet the requirements of aerobic metabolism, which will be at least 100 times greater. The difficulty increases, of course, with increasing activity. Thus it has been calculated that the transport of oxygen by diffusion in the tracheae of a flying insect will cease to meet its needs by the time it reaches a weight of 0.1 g. In practice, active insects reach much higher weights than this, being able to do so because the movement of oxygen is aided by ventilation movements of the body. A further helpful device is the expansion of the tracheae into air sacs, which, by increasing the volume of the system, increase the volume of air which can be changed (tidal air). Ultimately, however, tracheal respiration has restricted the size of insects; fortunately, we may well think. Tracheae can also fulfil other functions; serving, for example, as air stores, or as hydrostatic organs in aquatic insects, and aiding the energetics of flight by decreasing the specific gravity. It will be noted that there is some analogy here with the air sacs of birds.

Although tracheae are primarily an adaptation to terrestrial life, they have not prevented insects from returning to the water. The same is true, of course, of the lungs of mammals, which also have not prevented the secondary evolution of aquatic life. Nature is resourceful in these matters. Some aquatic insects simply continue to breathe air through their spiracles, returning to the surface for this, and sometimes carrying with them under the water a store of air attached to their body. Another device, independently evolved on many occasions, is the use of tracheal gills, which are essentially an accumulation of tracheae immediately below the body surface. Commonly these are associated with an air film, held in place by a complex of hairs or of tubercles and struts. This device, called a plastron, facilitates respiratory exchange by virtue of the large air-water interface which it provides. Moreover, oxygen can diffuse from the water into the air store which, because of its structural basis, tends to retain the same volume, thus permitting a prolonged stay under water. It provides an instructive comparison with the lung of aquatic pulmonate molluscs, in which, as we have seen, the air store needs frequent replenishment.

6 Salts and Water

Ionic Composition

The dependence of life upon water, and more especially the influence of this upon the lives of animals, needs to be considered in an evolutionary context. The fluids of marine invertebrates have an osmotic pressure that is often the same as that of sea water; which is one of the reasons for believing that life originated in the oceans, and that cells and tissues have continued to be dependent upon the conditions that existed there. Nevertheless, the ionic content of body fluids differs in some respects quite sharply from that of the sea, even in the lowliest animals (Table 6.1). At one time it was supposed that the present composition of those fluids reflected the conditions in primordial oceans, the composition of sea water having subsequently changed. This view, however, cannot be sustained, for there is no evidence that sea water has changed in the way required. The fact is that the ionic composition of the body fluids is not an evolutionary relic, but is maintained by active processes, requiring the expenditure of energy. Dynamic equilibria are thus established, and these vary from group to group.

Table 6.1 Ionic composition of the body fluids of some marine invertebrates.

Animal	Concentrations in plasma or coelomic fluid as percentage of concentration in body fluid dialyzed against sea water, and thus in passive equilibrium with it.					
	Na	K	Ca	Mg	Cs	SO$_4$
Coelenterate						
Aurelia aurita (jellyfish)	99	106	96	97	104	47
Echinodermata						
Marthasterias glacialis (starfish)	100	111	101	98	101	100
Annelida						
Arenicola marina (lugworm)	100	104	100	100	100	92
Arthropoda						
Maia squinado (crab)	100	125	122	81	102	66
Carcinus maenas (crab)	110	118	108	34	104	61
Mollusca						
Pecten maximus (scallop)	100	130	103	97	100	97

Evidence for this comes from studies that have exploited remarkable advances in methodology achieved in recent decades. Amongst these are microchemical procedures for determining the composition of minute samples of fluid, used in conjunction with techniques, such as the micropuncture of delicate tubules, for the removal of such samples. Perhaps the most important of all, however, has been the use of isotopes, and particularly radioactive ones, which have made it possible to study with precision the two-way movement of labelled material across biological membranes.

It is now evident that the ionic composition of body fluids, although deriving from that of the marine environment in which life first evolved, has been subject from an early stage to adaptive modifications. The significance of these modifications is not clear, although they are presumably related to the known importance of certain ions in various biological processes, such as excitation (p. 205) and contraction. What is certain, however, is that animals are confronted with the need to maintain the appropriate ionic composition in environments which do not always favour this.

An important property of biological fluids, and of the media in which they function, is osmotic pressure. We have already seen examples of this in considering some of the water relations of plants, in which context the term osmotic potential is preferred. (The two are numerically equal, but osmotic pressure is positive and osmotic potential negative.) Osmotic pressure, which is the pressure required to prevent water moving into an aqueous solution through a membrane permeable only to the water, depends upon the number (and not the kinds) of particles (ions and undissociated molecules) present in unit volume of the solution. Since their presence lowers the freezing point of the water by an amount proportional to their concentration, osmotic pressure is commonly measured and expressed as freezing point depression (Δ). Other units, however, are also used, and there is no uniformity in this. For example, the osmotic concentration may be expressed as a percentage of the concentration of sea water, or as millimoles of sodium chloride per litre. Another much favoured way is to express the concentration in terms of molality, which is the number of moles per litre. The unit in this case is one osmole, which is a measure of the number of particles in one gram molecule of a substance which does not dissociate in solution. Sodium chloride dissociates almost completely in aqueous solution; a solution containing one gram molecule kg^{-1} will thus have an osmolality of nearly 2 Osm kg^{-1}. In biological data it is usual to use the more convenient milli-osmole (mOsm), which is 1/1000 of an osmole.

Invertebrates

Animal cells, like those of plants, are profoundly affected by the osmotic pressure of the medium surrounding them, for changes in this result in water or ionic fluxes which disturb the internal osmotic pressure of the cells, and the functioning of their metabolic pathways. Conditions in the sea are so constant that this presents no problem for the many marine invertebrates that can survive without being able to regulate their osmotic pressure. A typical example of this is the spider crab, *Maia squinado*, (common on sea shores, where it protectively dresses itself with pieces of weed; Fig. 6.2). The internal osmotic pressure of this crab varies in close conformity with variations in that of the medium (Fig. 6.1). It remains isosmotic with the medium, and is termed an osmo-conformer, a condition which it can readily sustain because of the osmotic constancy of the medium in which it normally lives.

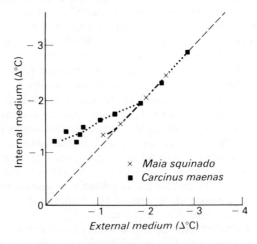

Fig. 6.1 Osmotic pressures (expressed as depressions of freezing points) of the blood of the crabs *Maia squinado* and *Carcinus maenas* as a function of the external medium. (From Barrington, 1968.)

Conditions cannot everywhere be as constant as this. Rainfall, evaporation, tidal movements, and the flow of fresh water must all influence the ambient osmotic pressure on the sea shore and in estuaries. These conditions, however, characterize the routes by which animals must have penetrated into fresh water and onto dry land. Moreover, animals survive them successfully at the present day. There must, therefore, be ways of overcoming the resulting osmotic stresses. One device is for the animal to isolate itself from the environment for short periods. The limpet does this, when the falling tide leaves it exposed, by adhering firmly to rock surfaces, while the mussel (*Mytilus*) achieves the same result by closing the valve of its bivalved shell (its habit of living in large clusters further aids the isolation of individuals in the centre of the mass). There is, however, another solution to the problem, of much greater evolutionary potentiality. This is the regulation of intracellular osmotic pressure in some degree of conformity with ambient changes. Animals which can do this are called osmoregulators. Because they can tolerate a range of osmotic pressures, they are termed euryhaline (*eurys*, wide); those which cannot do so are termed stenohaline (*stenos*, narrow).

Regulation of intracellular osmotic pressure depends upon this being maintained only in part by inorganic ions (about 50 % in some euryhaline molluscs, for example, as compared with 99 % in the extracellular fluid); the remainder is maintained by organic compounds, including amino acids, taurine and glycine-betaine. Euryhaline molluscs placed in dilute sea water have a lower intracellular level of these than when they are in normal sea water. Probably the metabolism of these compounds is regulated in response to changes in the osmotic pressure brought about by changes in the osmotic pressure of the sea water.

The same device is used by the lugworm, *Arenicola marina* (Fig. 4.11), an animal which has only limited powers of osmoregulation. It swells in sea water of reduced salinity, as a result of the osmotic entry of water, with consequent dilution of its body fluids, and it cannot survive at salinities below 12‰. Over its survival range, however, it lowers the intracellular osmotic pressure by reducing the concentration of certain amino acids, including alanine and glycine, which decline from their normal level of 427 nmole l^{-1} to 180 nmole l^{-1}.

An instructive group for the study of these problems is the class Crustacea. Few of them have succeeded in completely leaving the sea, but some have adapted themselves efficiently to littoral and estuarine conditions, and we shall see that the devices used present some interesting parallels with the vertebrates. A much-studied example is the shore crab, *Carcinus*, an osmoregulator which is found in the littoral zone and in the brackish water of estuaries (Fig. 6.2). Regulation (Fig. 6.1) is achieved in part by intracellular osmotic adjustments similar to those just outlined, although inorganic ions are involved as well as amino acids. Potassium, for example, remains constant as the osmotic pressure of the surrounding water falls, but the concentrations of sodium and chloride are lowered. Another factor is surface permeability, which

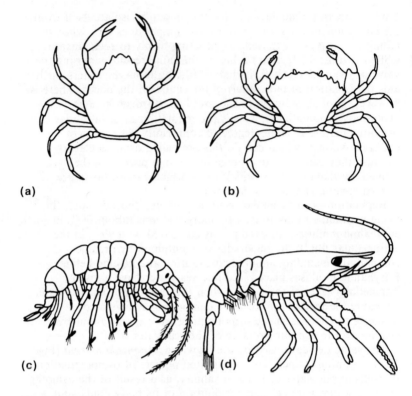

Fig. 6.2 (a) *Maia*, (b) *Carcinus*, (c) *Gammarus* and (d) *Astacus*. Not to scale.

is lower in brackish-water crustaceans than in marine ones; the former are therefore better protected from the stresses of water flux. The effect is exemplified by a comparison of *Carcinus* with the edible crab *Cancer*, which is wholly marine. These animals discharge urine, secreted by the antennal glands (often called the kidneys), through a pair of excretory pores, situated at the bases of the second antennae. If the pores are blocked, the urine accumulates and the animal increases in weight, the increase reflecting the entry of water into the body through the permeable surface. The increase is significantly less in *Carcinus* than in *Cancer* showing that the latter has the more permeable surface.

Finally, *Carcinus* makes use of active uptake of ions as a reaction to a fall in external osmotic pressure, a device that can be demonstrated by the use of radioactive sodium to show the influx and efflux of the ion. In normal sea water there is some loss of sodium in the urine, and by diffusion through the body surface, but this is replaced by passive inwards diffusion. (It will be noted that radioisotope studies reveal movements in both directions; these could not be detected merely by measuring sodium levels, for this would only show the balance resulting from efflux and influx.) With continued fall in external osmotic pressure, there comes a point at which active uptake of sodium is initiated, so that the animal can now maintain a level of the ion higher than that in the external medium. There is, however, a limit at which blood sodium levels begin to fall, so that under experimental conditions the animal can only partially counteract continued external decline. Nevertheless, its capacity is sufficient to meet its normal need in brackish water.

The weakness in the complex of adaptations operated by *Carcinus* is that its urine is isosmotic with the blood. So it is also in other freshwater crabs, such as *Eriocheir*, the woolly-handed crab of Asiatic rivers and rice paddies, which has to return to salt water for breeding. There is thus a substantial loss of ions from these animals, and this has to be made good by active uptake, with consequent expenditure of energy. Hence the value of an adaptation found in freshwater crayfish such as *Astacus* (Fig. 6.2), as well as in the freshwater mussel *Anodonta*. This is the production of a hypo-osmotic urine (that is, with an osmotic pressure lower than that of the blood), made possible by the kidney reabsorbing ions from the excretory fluid passing through it. Some ions are still lost, but to a much less extent than in *Carcinus*, with a resulting economy of energy. The economy is increased by a reduction in internal osmotic pressure as compared with that of the crabs mentioned ($\Delta = -0.6°$ to $-0.8°$ in *Astacus* as compared with -1.18 in *Eriocheir*). This reduction, made possible because the tissues are adapted to function at these low levels, results in a smaller loss of ions by diffusion to the ambient medium, because of the reduced osmotic gradient. Less energy has therefore to be expended in ion uptake.

The production of hypo-osmotic urine by *Astacus* is correlated with the presence in the antennal gland of an additional long tubule which connects the greenish labyrinth with the bladder, and which is not found in the antennal glands of marine crustaceans. This region, while not the sole site of ion absorption (an example of the detail revealed by micropuncture studies), nevertheless makes a contribution to it. A similar adaptation has been independently evolved in some of the species of the amphipod *Gammarus* (Fig. 6.2). *G. locusta* is a marine form, with only a short tubule. *G. duebeni* lives in brackish water and has a longer tubule, while *G. pulex*, which lives in fresh water, has a still longer one. Evidence for its functional significance is that the last two species can both produce a hypo-osmotic urine.

Cyclostomes and Fish

The vertebrates provide an excellent illustration in this context of the exploitation of similar devices by unrelated groups in response to similar needs. There is good geological evidence that these animals originated in the sea, probably in inshore waters, and then moved into fresh water. Some of them later became terrestrial, while others returned to the sea. A consideration of their osmotic relationships shows that their movement into fresh water must have involved the development of adaptations similar to those just discussed. One might assume, by analogy with the invertebrates, that the vertebrate stock was at one time iso-osmotic with sea water, and this is indeed so in the myxinoid cyclostomes (e.g. *Myxine*, the hagfish). The class Cyclostomata belongs to the Agnatha, a group of jawless vertebrates which preceded the jawed vertebrates. Modern cyclostomes, despite the many specializations to be expected in a group which has survived so long, still preserve primitive characteristics as well, exemplified by their haemoglobins. The iso-osmotic condition of the myxinoids is almost certainly one of these.

The other group of surviving cyclostomes, the lampreys (Petromyzontia), spawn in fresh water (Fig. 6.3), producing larvae which may, after metamorphosis, pass to the sea before returning to spawn. It is doubtless in correlation with this freshwater phase in their life histories that their internal osmotic pressure, like that of the crayfish, is lowered below that of sea water. But here the adaptation has been of fundamental importance to the whole group, for all vertebrates, with the sole exception of the myxinoids, and apart from the special case of elasmobranch fish (to be mentioned later), have this lowered osmotic pressure, although the general balance of their ions conforms broadly with that already outlined.

This osmotic pattern, which must have been established early in vertebrate history, still leaves problems, the resolution of which involves the action of the gills and of the kidneys (Fig. 6.4). Freshwater teleost (bony) fish (and the lampreys in their freshwater phase) suffer from an osmotic influx of water, for their osmotic pressure is well above that of freshwater. This influx is removed through the kidney in a hypo-osmotic urine, the consequent loss of ions being compensated by active uptake through the gills. Marine teleost fish, which evolved from freshwater ancestors by a secondary return to the sea, retain the osmotic pressure of those ancestors, which is thus below that of sea water. They therefore suffer from an osmotic outflow of water, and they compensate for this by drinking sea water. The monovalent ions contained in this are absorbed by the intestine and lost through the gills, while the divalent ions, which are mostly not absorbed, are lost in the faeces. The lamprey,

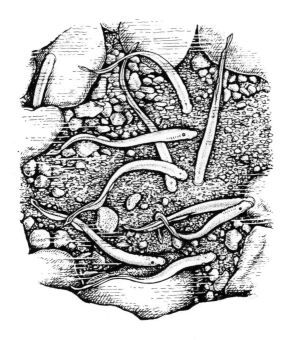

Fig. 6.3 Spawning lampreys seen in their nest. (From Young, 1962; after Gage (1961), *New Scientist*, **27**.)

which migrates into the sea during one phase of its life history, changes the method of regulation to conform to the medium. So also do migratory teleost fish such as the salmon and eel.

Elasmobranch (cartilaginous) fish (Fig. 6.5) constitute a special case. These are mainly marine fish, which for no obvious reason, but exemplifying the unpredictability and diversification of the living world, differ from other fish in raising their osmotic pressure to the level of sea water, or slightly above it, by increasing the urea content of their blood to levels which would be fatal to other vertebrates. A high level of trimethylamine oxide also contributes to this elevation. This nitrogenous substance is excreted in large amounts by marine teleosts, but in elasmobranchs it is largely reabsorbed in the kidney, together with urea. Because of this adaptation, the elasmobranchs do not suffer from the osmotic outflow of water to which marine teleosts are subject, and are therefore able to conserve energy by not having to rectify this. However, although the osmotic pressure of the blood is high, the inorganic ionic concentrations are low, which means that there is inward diffusion of ions. We have seen that in teleosts the gills

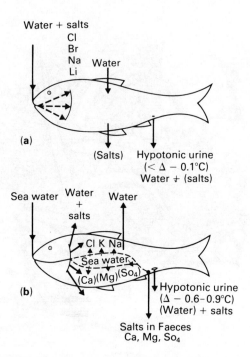

Fig. 6.4 (a) Osmotic regulation in *freshwater* teleost fishes. (b) Osmotic regulation in *marine* teleost fishes. (From Barrington, 1968.)

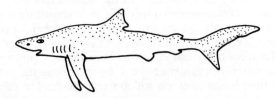

Fig. 6.5 *Prionace glauca*, the great blue shark, a widely distributed fish-eating and scavenging elasmobranch.

provide for active transport of ions out of the body, and it is now believed that efflux of sodium and chloride takes place across the gills of elasmobranchs as well, although it used to be thought that the rectal gland was mainly responsible for this. The urine, which is hypo-osmotic, removes divalent ions.

The osmoregulatory mechanism of elasmobranchs largely commits them to a marine habitat, but some do penetrate into freshwater rivers and lakes in the tropics. These species have blood urea levels lower than those of their marine relatives, although the urea is never entirely lost. In this respect they are at some disadvantage compared with the teleosts. Certainly the latter group are far more successful than elasmobranchs, as judged by diversification, numbers and distribution, but many other features of organization also contribute to their success.

Non-mammalian Tetrapods

Osmoregulation in the terrestrial vertebrates (tetrapods) is complicated by their need to conserve water. This aspect, which, of course, influences the organization of terrestrial invertebrates as well, will be dealt with later, and in the present context a few general points will suffice.

Amphibians, as we have already seen, are greatly restricted by their use of a permeable skin as a respiratory organ. They are unable to control evaporative water loss through this, and are therefore compelled, in general, to remain in moist habitats. In adaptation to this, however, they can control their rate of production of urine; this is greatly reduced when they are out of water, salt losses being replaced from their food. When they are living in water, there is an osmotic influx of water through the skin, and urine production is greatly increased. The increased loss of salt is now compensated for by active uptake of sodium through the skin.

Those reptiles which live on land have a largely impermeable skin, and in consequence of this advance they enjoy greater freedom of movement than do amphibians; there is, however, considerable evaporative loss from their lungs. Rates of urine production are low, but the urine is still hypo-osmotic, like that of lower vertebrates. Water and ions are probably replaced largely through the food, although reptiles do also drink. As we shall see, the low rate of urine production is made possible because nitrogen is excreted as uric acid, instead of urea (as in amphibians). The advantage of this is that the low solubility of uric acid allows it to be voided as a semi-solid material, and renders unnecessary the production of a large volume of fluid to remove it. Aquatic reptiles are somewhat differently organized. They may excrete urea, and some have a very permeable skin. The caiman, for example, is estimated to lose water by evaporation at a rate as great as one-half that of amphibians, most of the loss taking place through the skin. Birds, like reptiles, have a largely impermeable skin, but their respiratory loss of water is very high, partly because of their high rate of metabolism, and partly because of its involvement in the control of body temperature, another aspect to be considered later. However, urine production is low, because, like reptiles, they can excrete nitrogen as uric acid; moreover, some birds, unlike reptiles, can produce a urine that is hyper-osmotic to the plasma.

Marine birds face a special problem, as do also certain marine reptiles, such as the green turtle, *Chelonia mydas*. It is uncertain whether these animals normally drink sea water, although the albatross is thought to do so. But whether water is obtained in this way or in the food, these animals must absorb a substantial salt load which cannot be wholly eliminated through the kidney because the urine is hypo-osmotic to the plasma. They use instead a pair of glands (nasal glands, salt glands), situated above the orbits, which secrete a fluid containing sodium chloride at a level hypertonic to the plasma. Release of the fluid is readily induced in duck, for example, by administering a salt load. These glands seemed at one time to present an interesting analogy to the rectal glands of elasmobranchs, but the importance of the latter in osmoregulation, as we have seen, is now less certain, and the salinity of their secretion is appreciably lower.

Salt glands are especially characteristic of birds living in a marine environment, yet the savannah sparrow, which inhabits salt marshes, and which could certainly make use of this adaptation, does not possess them. Its condition, nevertheless, is not so disadvantaged as this might suggest, for it is one of the birds that can produce a hyper-osmotic urine, with a salt concentration similar to that of the salt gland secretion. Such are the unpredictable vagaries of animal adaptation.

Mammals

The contribution of the vertebrate kidney to the regulation of water and ions is best understood in mammals (Fig. 6.6), for its importance in medical physiology has concentrated a great deal of research effort upon it. As in other vertebrates, it is formed of unit kidney tubules (nephrons), but the mammalian nephron includes a characteristic segment called Henle's loop, with a descending and an ascending limb; most of the medulla of the organ is formed by a great concentration of these loops. This segment, represented also in birds but in a different form, is largely responsible for the mammalian kidney being able to produce a hyper-osmotic urine, but the nephron is further adapted, in its structure and in the way in which its action is regulated, to vary the composition of the urine as the fluid passes along it.

Production of the urine begins in the renal corpuscle. Here, with

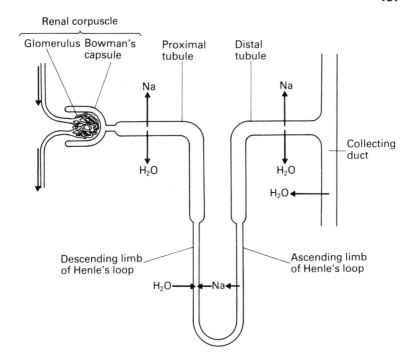

Fig. 6.6 Mechanism of urine formation in the mammal.

blood pressure as the driving force, an ultrafiltrate, resembling the blood plasma in composition except for the absence of protein, passes from the vessels of the glomerulus into the cavity of Bowman's capsule and on into the first part of the nephron (proximal tubule). Here its composition is modified by active uptake of sodium, and by the passive movement of water which accompanies this. Other substances, including glucose, are also returned to the blood stream at this stage.

The key to the production of a hypertonic urine is, however, further active uptake of sodium from the second (ascending) limb of Henle's loop. The close apposition of the blood vessels and of the two limbs of the loop permits recycling of this sodium through the tissue spaces of the medulla into the descending limb, with some accompanying movement of water, but the presence of the sodium in the medulla creates up to a four-fold rise in osmotic pressure. The existence of this concentrated 'brine bath' results in water being withdrawn along the osmotic gradient from the urine as this passes along the distal tubule and along the collecting duct into which the tubule discharges. The amount of water movement, however, both here and in the ascending limb of Henle's loop, can be regulated by a hormone (vasopressin, antidiuretic hormone) which influences cell permeability in response to physiological need. Movement of sodium can also be regulated, in the distal tubule, by the hormone aldosterone, produced by the adrenal cortex.

The final result of this sequence of events, described here in simplified outline, is the production by mammals of a urine which can be delicately adjusted in its composition to environmental pressures, but which is, in general, more concentrated than that of birds. Probably this is the route by which salt loads are eliminated in those mammals which have returned to the sea, although the mechanisms of salt and water balance in some of them is not well understood. They do not, however, possess salt glands. Two other differences from birds may be noted. Many mammals possess sweat glands (man and the horse, for example) which release a secretion of water and salts; these are part of the temperature-controlling mechanism (Chapter 7). A second difference is that mammals eliminate nitrogenous waste as urea instead of as uric acid, and this must be removed in solution, in contrast to the pasty material which birds can evacuate. The difference is related to the mode of life of the embryo. Urea can be readily removed from the circulation of the foetal mammal through the maternal blood stream (p. 196), whereas in birds the deposition of the relatively insoluble uric acid inside the egg is a more appropriate way of removing nitrogenous waste,

7 Life, Heat and Temperature

Heat transfer

In considering the biological importance of solar radiation we have so far dealt only with its photochemical effects. We must now consider its thermal activity, which extends, beyond the limits of the visible spectrum, from the ultra-violet to the infra-red. Within this range it provides for living organisms an input of energy in the form of heat; an input of great biological importance because it brings about molecular agitation and thus increases chemical activity. We shall see that organisms are subjected to heat transfer along a number of pathways (Fig. 7.1), the direction of the transfer being determined by the temperature of the bodies concerned. This is a function of molecular agitation, and is a measure of the amount of heat energy that they contain. Heat will flow from a body at higher temperature to one at lower temperature; when they are at the same temperature no heat will flow, and they are then said to be in thermal equilibrium.

Another factor important in heat transfer is the specific heat capacity of the bodies involved, for this determines the amount of energy that is transferred. The specific heat capacity of a system can be expressed as:

$$\text{heat capacity} = \frac{d\text{H}}{m\,d\text{t}}$$

where $d\text{H}$ is the change in heat of a system, m its mass, and $d\text{t}$ the difference in temperature between it and its surroundings. The value varies with temperature, but for practical purposes the specific heat capacity of water can be taken as $4.184\ \text{J}\,\text{g}^{-1}\,^\circ\text{C}^{-1}$; values for animal tissues, which have a high water content, are taken to lie between 2.9 to 3.8 J. These values are high in relation to those of many common materials, such as lead (0.125), copper (0.375) and glass (0.835), and we shall see later that the high specific heat of water (a result of the large amount of energy needed to break the hydrogen bonding of its molecules) is of great biological importance.

The relative importance of the biologically significant pathways for heat transfer depends primarily upon whether the organisms concerned are living in water or in air (Fig. 7.1). Heat can be

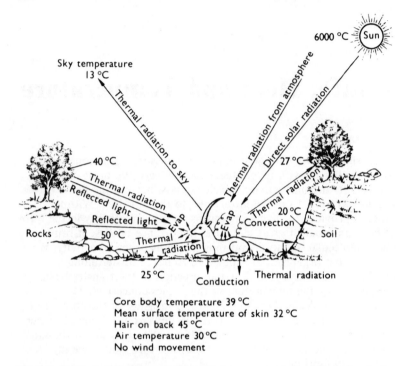

Fig. 7.1 Diagram of the energy exchanges between a mammal and its environment under moderately warm conditions. (From Gordon *et al.* (1968). *Animal Function: principles and adaptations.* Macmillan, New York and London.)

absorbed from short-wave solar radiation, direct or reflected, and can be transferred from body surfaces by long-wave terrestrial radiation in the range 3–100 μm. Another mode of heat transfer is by conduction, which is defined as occurring between two bodies that are in contact but that are not moving in relation to each other. Materials of the environment differ greatly in their thermal conductivity, since this is due to energy exchanges between molecules in the conducting medium. The value for air is low, that for water somewhat higher, but both are very low in comparison with metals. These differences, too, can be of great biological importance.

Heat can also be transferred by convection through air and water, and, of course, through other gases and liquids. This depends upon the setting up of convection currents in the medium, as a result of its density varying with temperature. Finally, large amounts of heat can be transferred by the evaporation of water, because of the large amount of energy needed to convert it from liquid to gas (another consequence of its hydrogen bonding). All of these pathways are seen at their most complex on land (Fig. 7.1). We shall consider shortly how the challenge that they represent has been exploited and overcome, both on land and in water.

In general, biochemical reactions can only function as components of biological systems over a limited temperature range, extending upwards to about 50°C and downwards to just below the freezing point of pure water, the upper limit being supposedly imposed by the progressive denaturation (inactivation) of enzymes with increasing temperature. There are exceptions to this generalization, for life can sometimes be maintained below and above these extremes, but individual species are commonly adapted to survive and to function with optimal efficiency within this range at what is called their preferred temperature, which is a characteristic of the species. Equally characteristic is the limiting or lethal temperature; this is the temperature beyond which survival is impossible if the organism cannot escape from it.

These considerations are well illustrated by micro-organisms (fungi and bacteria). Most have temperature optima between 20 and 45°C; they are called mesophiles. Some, called psychrophiles (*psychros*, cold), have optima below 20°; these include organisms adapted for life in cold oceanic waters, but some can cause spoilage in cold food stores. Others, called thermophiles, have optima above 45°, and can flourish (in compost heaps, for example) at temperatures which would kill most forms of life. An example is *Mucor pusillus*, which has been identified in stores of rotting maize at 58°C. Some can even survive in hot springs, where they form the basis of a self-contained ecosystem.

Comparable influences of temperature are familiar in higher plants. Thus, to give only one illustration, the temperature optimum for the growth of maize is 30–35°C, while that for winter wheat is 20–25°C, the difference reflecting the ambient temperatures of the normal environments of these plants. But the response to temperature may be adaptively related to different phases of the life cycle. Some plants (annuals) will begin their growth in the spring and flower later in the same year. Others (biennials) produce only vegetative growth during the first year, flowering (which is often promoted also by long days) occurring during the second year after exposure to winter cold; sugar beet is an example. This cold treatment (vernalization) may be essential in some species if flowering is to occur, while in others (e.g. winter rye) the experimental application of it may simply accelerate the onset of the flowering. Vernalization is essential for henbane (*Hyoscyamus niger*), which remains permanently vegetative if it is kept at too high a temperature during the winter. Then again, the growth of plants may be improved by an alternation of higher day temperatures with lower nocturnal ones, and it is likely that the preferred optimum will change during the life history of the plant. Underlying these responses is the fundamental principle that life is the resultant of an interacting complex of biochemical reactions, which must be in dynamic balance with each other. Individual reactions will be affected in different ways by different environmental factors, and will vary in their relative importance at different phases of the life cycle.

Similar principles apply to animals. Many live within quite narrow temperature limits, and with preferred optima that are adaptively characteristics of individual species and developmental stages. But there is one factor which greatly complicates the temperature relationships of animals: their mobility. We shall see later how this makes it possible for animals to seek out small scale environments (microhabitats) which can provide conditions that are significantly different from those in the larger environments of which they form a part.

Temperature Coefficients and Characteristics

The biological importance of temperature for living organisms follows from its influence upon the rates of biochemical reactions. This influence is commonly expressed in terms of a factor called the temperature coefficient (Q_{10}), which is the rate by which a reaction is increased for a rise in temperature of $10°C$. The coefficient is derived from the equation:

$$Q_{10} = \left(\frac{K_1}{K_2}\right)^{10/(t_1 - t_2)}$$

where K_1 and K_2 are the velocity constants at temperatures t_1 and t_2. Since K_1 and K_2 are proportional to the corresponding reaction rates (V_1 and V_2), the latter are often used instead. According to a generalization due to van't Hoff, the Q_{10} value for most biochemical (enzymatic) reactions lies between 2 and 3, a value of 2.5 indicating an increase in reaction rate of 9.6% for each $°C$ rise in temperature. Values for physical processes, such as diffusion, are below 1.5, and this should be borne in mind when evaluating biological data. So also should the variation of values with temperature, especially in biological systems when these are approaching the extremes of temperature tolerance. Thus the temperature range to which Q_{10} values apply should always be stated.

The acceleration of an enzymatic reaction by an increase in temperature is due in part to increased molecular agitation, but this cannot provide the whole explanation, for the increase is greater than could be accounted for in that way. The nature of the relationship was investigated at the turn of the century by Arrhenius, who arrived at an initially empirical formulation which can be stated as $k = Ae^{-E/RT}$ or, in logarithmic form, as $\ln k = \ln A - \dfrac{E}{RT}$. In these equations, A is a constant, k is the reaction velocity constant, T the absolute temperature, R the gas constant ($8.3\ \mathrm{J°C^{-1}\ mol^{-1}}$), and E (often written μ) another constant termed the activation energy of the reaction.

The Arrhenius equation is the key to explaining the size of the temperature coefficient of biochemical reactions. In principle, it is supposed that the reactions depend upon the intermediate formation of high energy (activated) molecular complexes of the reactants, and that the formation of these transition states increases rapidly with rising temperature. Reaction rates are thus accelerated by far more than would be possible if molecular agitation were the only factor involved. It will be noted from the equation that there is an exponential relationship between the rate constant (k) and the absolute temperature (T). It is because of this that a relatively small change in temperature brings about such a large change in the velocity of the reaction.

Activation energy values (also known as temperature characteristics, or Arrhenius constants), which are expressed as calories or joules, are constant over at least a limited temperature range. They may be determined, in accordance with the Arrhenius equation, by plotting $\ln k$ against the reciprocal of the absolute temperature. This gives a straight line with a slope equal to A/R with natural logarithms, or $A/4.58$ ($= A/2.30\,R$) with logarithms to the base ten. (Further details and discussion of these matters will be found in Hardy, 1972, and in Morris, 1974.)

The values thereby obtained are characteristic of particular reactions, and also for particular biological processes. Thus the chirping of crickets has a value of about 51 000 J, while respiratory and cardiac rhythms of many species have values of about 69 900 J. Such complex biological processes necessarily depend upon a sequence of individual reactions. It has therefore been postulated that the temperature characteristic of a particular process would be determined by the characteristic of the slowest reaction involved, which could thus be termed the master reaction, and that there might be a switch from one master reaction to another with change of temperature. No doubt there is an element of truth in this, for the individual reactions may well differ considerably in their temperature characteristics, and hence in their rate-limiting effects at different temperatures. But the interaction of these reactions must also be very complex, while physical processes with quite different characteristics (see above) are likely also to be involved. It is not surprising, therefore, that extensive research aimed at identifying individual master reactions has never attained this objective.

Poikilothermy and Homeothermy

The unique physical properties of water, largely due to the hydrogen bonding mentioned earlier, are of immense biological importance for the temperature relationships of both plants and animals. Its high specific heat, enabling it to take up large amounts of heat with relatively small changes of temperature, creates for aquatic organisms a degree of thermostability which is not attainable on land. The temperatures of open waters range from about $-2°$ to $+40°C$, a major contrast with conditions on land, where substratum temperatures ranging from about $-65°$ to $+70°C$ have been recorded. Because of the high latent heats of melting and evaporation of water, a high input of energy is needed to change its physical state, and this contributes to its thermostability. Moreover, it has an exceptionally high freezing point, while its behaviour when it does freeze is unique, for, although its freezing point is at $0°C$, it attains maximum density at $4°C$ and therefore tends to sink. As a result, ice forms at the surface of the water, while immediately beneath this ice layer there is water which remains fluid and thus permits (in conjunction with adaptations to be mentioned later) the maintenance of life below the ice.

The benefits derived from these properties are not restricted to those organisms that actually live in water; they also affect life on land. The thermal stability of water has a moderating influence upon the temperatures of land adjacent to it, as is readily apparent on comparing the climates of islands and continents, or seaboards and inland areas. Moreover, its evaporation carries large amounts of heat into the upper air, and this heat is distributed over the land when the water vapour falls as rain. Harsh though the terrestrial climate may sometimes be, it would be much more rigorous were it not for the influence of water, and we shall see more of this when we consider the problems that face organisms that have to survive in arid climates.

Plants are unable to regulate their temperature. Exposed as they are, and typically immobile, their internal temperature varies little from the ambient, whether in water or on land, although temperatures within the trunks of trees exposed to the sun may rise substantially above that of the air. The heat generated by metabolism is sometimes turned to biological advantage, the wild arum (*Arum italicum*), being a remarkable example, for a substantial rise in temperature within its inflorescence serves to attract the thermophilic flies which effect pollination. Tolerance and hardiness, however, must be the main defences of plants against temperature stresses. These genetically determined properties lend themselves well to exploitation by man for the production of strains capable of survival and high productivity under rigorous conditions.

The same qualities are important for animals, but for them the possibilities are much more complex. Here it is necessary to clarify some points of terminology. A distinction is often made between cold-blooded animals and warm-blooded ones, but this terminology is quite unsatisfactory, and not only because it is based upon our naive responses to handling vertebrates. It amounts to saying that the temperature at the surface of cold-blooded animals is lower than that of our fingers at the moment when we touch them, and at the site of our touch. But their surface temperature may be very different at other times, and so also may be the temperature in the interior of their bodies. This is referred to as the inner or core temperature, conveniently defined (at least in vertebrates) as the temperature at one inch or more below the body surface.

For these reasons the preferred terms in this context are poikilothermic and homeothermic (or homoiothermic). By poikilothermic (*poikilos*, various) is meant that the temperature of the body is variable, but whether or not the body is cold depends upon circumstances that we consider later. By homeothermic (*homoios*, like) is meant that the temperature of the body, like that of our own, is more or less constant, although this does not exclude the possibility of some variation, especially at the body surface.

An important concept is involved in the terminology. Metabolism generates heat, but it is characteristic of poikilothermic animals (as also of plants) that this heat is normally lost to the environment, often as rapidly as it is produced. Because of this, the body temperature of poikilotherms is determined primarily by the amount of heat that they take up from their environment. Such animals are therefore termed ectothermic. Birds and mammals, however, are so constructed that much of the heat that they generate is retained within their bodies. It is upon this that their high body temperature depends, the effect being greatly enhanced by them having in general a much higher rate of metabolism than that of poikilotherms. They are therefore termed endothermic. There are, however, exceptions which do not conform to this simple dichotomy. Certain animals, both vertebrate and invertebrate, can sometimes attain body temperatures substantially above that of the environment (p. 142); often this is only for brief periods, but reptiles can maintain these elevated temperatures at a constant level for a remarkably long time (p. 145). It is convenient to apply to such animals the term heterothermic. With this terminology as a basis,

we may now examine the thermal relationships of animals with the
different types of habitat that they have exploited, beginning first
with those that live in water.

Acclimation and Acclimatization

Poikilothermal animals living in water, like the aquatic flora,
cannot, as a general rule, maintain a temperature different from the
ambient. They cannot lower their temperature below that of the
water because the only way in which animals can achieve a
temperature below the ambient is to lose heat by evaporation. This
is a device well exploited by terrestrial animals, but it is obviously
impossible for aquatic ones. It is also impossible, with certain
exceptions to be mentioned below, for aquatic animals to raise their
temperatures above the ambient, for their heat of metabolism is lost
to the water as quickly as it is generated. This is due in part to the
high heat capacity of water, but a contributory factor is that many
animals must maintain currents of water around the body for
respiration and, often, for filter feeding as well, both of which
processes greatly favour heat exchange. This is further promoted by
the relatively low oxygen content of water (about 1 part in 100),
which makes it necessary for large amounts of it to be passed over
the respiratory surfaces, while filter feeders may need to move large
amounts in order to secure an adequate food supply. Certain
ecological consequences may be expected to follow from all of this.
The distribution of aquatic poikilotherms is closely influenced by
water temperatures, while individual species tend to be adapted for
life in preferred temperatures, and if they are free-swimming, to
have receptor and motor systems enabling them to respond to
temperature fluctuations. Seasonal migration can be an important
element in such responses, although, as we shall see later, other
factors are also involved in such movements.

Adaptation to preferred temperatures is certainly a well-marked
feature of animal life, but it is carried to extremes which seem at
first sight to raise a considerable problem. This arises directly from
the Arrhenius equations, which, as we have seen, show that the
velocities of metabolic reactions are related logarithmically to
temperature. If this were all, it would follow that with lowering of
temperature, animals would pass into an increasingly inactive state,
while the reverse would happen as temperatures rose, to a point at
which feeding capacity would be unable to meet increasing
nutritional demands. In reality, animals are able to adapt in ways
that enable them to overcome this problem, with the result that
many can maintain, at temperatures close to freezing, a level of
activity quite comparable with that of related forms living at

temperatures 20° or 30°C higher. The full significance of this phenomenon of adaptation to environmental temperature, which is termed temperature compensation, was well stated by Barcroft. 'It is clear', he wrote, 'that nature has learned so to exploit the biochemical situation as to escape from the tyranny of a simple application of the Arrhenius equation. She can manipulate living processes in such a way as to rule, and not to be ruled by, the obvious chemical situation. That is true at least over a wide range of temperature.'

The validity of this conclusion has been amply demonstrated by observation and experiment, described in the following terminology. In dealing with temperature compensation, a distinction has often been made between that shown in natural conditions and that induced in the laboratory. The latter, which usually involves adaptation to a single parameter (temperature alone, in this context) is termed acclimation. The more complex changes involved in the response to natural conditions, in which it is impossible to isolate completely the influence of any one factor, is termed acclimatization. Sometimes, however, these two terms are used interchangeably; the mode of use will then be clear from the context.

Earlier investigations of temperature compensation tended to be comparative studies of parameters of activity in related species, some living in colder conditions at higher latitudes, for example, and others in warmer conditions at lower latitudes. These showed that rates of activity differ far less than would be expected from the Arrhenius relationship; sometimes, indeed, they are almost the same. The point is well illustrated by the prawn *Pandalus*. *P. borealis* is an arctic form which cannot survive at aquarium temperatures higher than 11°C, while *P. montagui*, a closely related species from the English Channel, normally encounters maximum temperatures of around 15°C. One easily measured parameter of activity is the rate of beating of the respiratory appendages. Figure 7.2 shows that the rate in *P. borealis* at 7.5°C is the same as that of *P. montagui* at 12°C. This is contrary to what would be expected from a strict application of the Arrhenius relationship, and indicates some form of genetically based adaptation, secured by the action of natural selection, which enables both species to maintain optimum activity in their normal habitats.

The same type of adaptation can also occur in different populations of the same species (Fig. 7.2). *P. montagui* is found both in the English Channel and in colder Swedish waters. Comparison of the populations shows that the rate of beat is the same in both populations, despite the difference in the temperature of their habitats.

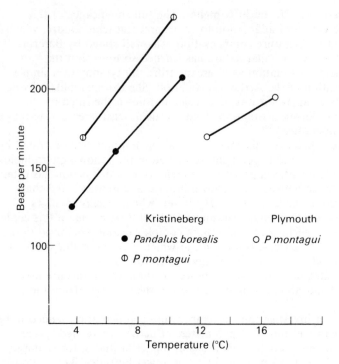

Fig. 7.2 Rates of movement of respiratory appendages of crustaceans at various temperatures.

A third possibility, also realized in practice, is that the same type of adaptation may operate within one species or population from season to season, thus enabling it to compensate, while remaining within the same habitat, for seasonal temperature changes, and allowing metabolic activity to be maintained at adequate levels throughout the year.

Temperature compensation is widespread in animals, but it must be sufficient to take one further illustration, in this case from the study of oxygen consumption at controlled temperatures in respiration chambers. Comparison of arctic forms of fish with tropical ones shows (Fig. 7.3) that the regression line relating oxygen consumption to temperature is displaced towards the lower temperatures in the arctic fish as compared with the tropical ones. The extent of the adaptation involved can be judged from the broken lines in Fig. 7.3, which show what the oxygen consumption of the tropical forms would be if extrapolated to 0°C. Actually, they could not survive at this temperature, but if they could, their

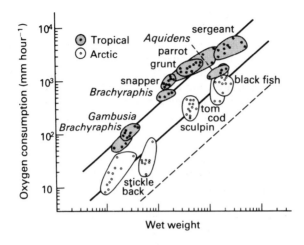

Fig. 7.3 Oxygen consumption in tropical and arctic fishes at their normal habitat temperature. The tropical forms extrapolated to 0°C would fall along the dotted line some thirty to forty times below the rates at 30°C. The arctic forms show a very marked relative adaptation to cold. (After Scholander *et al.* (1953). *Physiol. Zool.*, **26**, 72.)

consumption would be 30 to 40 times lower than their actual consumption at 30°C. So well adapted, however, are the arctic fish, for example, to the low temperatures of their normal habitat, that their oxygen consumption at 0°C is only 3–4 times lower than the consumption of the tropical forms at 30°C. Because it is indeed lower, it is sometimes supposed, here and in comparable cases, that the adaptation is incomplete or imperfect, but it is always unwise to attribute imperfection to nature, and this interpretation may be wrong. It is possible, for example, that metabolic demands may not be as great at lower temperatures. If this be so, the cold-acclimated organisms need not metabolize or respire as rapidly as their warm-acclimatized relatives.

What, now, can we say of the biochemical basis of temperature compensation? Undoubtedly any explanation must be sought in the genetically-determined properties of the intracellular enzymes which regulate metabolic reactions. We have already referred to the concept of 'activation energy', which is essentially a barrier that reactants must overcome before their reaction can take place. An enzyme facilitates a reaction by reducing the activation energy of that reaction, and this it does by transient binding of the reactants to itself. The rate of a reaction catalyzed in this way depends upon several factors. One of these is the temperature characteristic of the reaction; others are the concentrations of substrate, enzymes, and end products. The rate is directly proportional to the enzyme concentration, and also to the substrate concentration when this is low. At higher concentrations of substrate, the enzyme will eventually become saturated and the rate of reaction constant. It follows that temperature compensation could take place within the cell through regulation of rate-determining reactions, brought about by variations in either enzyme or substrate concentrations, or in both of these.

But there is also much scope for variation in the enzyme complement available for action within the cell. For example, an enzyme may be removed from the reacting system, or introduced into it. Such changes, termed respectively the repression or the induction of an enzyme, are brought about by direct action of the genetic mechanism of the cell, perhaps in response to qualitative or quantitative substrate changes (see p. 20).

Another possibility arises from the widespread existence of groups of structural variants of proteins, called allomers. The members of any one group of allomers are similar in function, but, because of their small differences in molecular structure, their actions are not actually identical. Enzyme variants of this type, known as isoenzymes (isozymes), provide scope for subtle modification of metabolic activities. They are exemplified by lactate dehydrogenase, which we have seen to function in glycolysis. Its molecule is a tetramer, composed of four units which, by association in different combinations, give rise to five isozymes. New isozymes of this enzyme appear in goldfish and trout during cold acclimation, having higher substrate affinities at lower temperatures. They thus counteract the tendency of many enzymes to show sharply reduced affinities at lower temperatures, with consequent reduced activity. Another example is the occurrence of two isozymes of cholinesterase in the brain of the trout (see p. 206). One of them, with maximum substrate affinity at 10°C, is present in warm-acclimated fish, the other, with maximum substrate affinity at 2°C, in cold-acclimated ones, while both are present in fish acclimated at an intermediate temperature of 12°C. No doubt this type of adaptation is one factor accounting for cold acclimation of fish in the laboratory taking from one to several weeks for completion; the time lapse is presumably the time needed for induction of new isozymes. It should be remembered, too, that many intracellular enzymes are bound to membranes. Probably, then, temperature acclimation will involve considerable restructuring of the architecture and composition of these membranes, facilitating efficient operation of the enzymes in the new intracellular environment.

Finally, studies of the genetics of populations in the field have shown that they carry a high degree of genetic diversity in their enzyme complement, and that the balance between the variant genes changes with the seasons and with other environmental influences. This is probably due to natural selection favouring different variants at different times, thereby improving the prospects for the survival of the population as a whole. The existence of a reserve of genetic diversity is thus a fundamentally important element in the mechanism of survival in a fluctuating environment.

Thermal Problems in Water

We have mentioned that life can only be maintained within a limited temperature range; towards the upper and lower limits of this range plants and animals are confronted by problems of survival. These, however, have been surmounted through adaptations evolved in consequence of the extreme pressure which always tends towards the colonization of every available ecological niche, in water as well as on land.

Some fish live in hot springs, but here they almost certainly rely upon behavioural responses which lead them to cooler microhabitats in the occupied area. Like most organisms, their upper limit for survival cannot be much above 42°C, although some insects can apparently survive at 50°C, while thermophilic bacteria and algae can live and even reproduce at 70°C.

These are rigorous conditions, and no less so is extreme cold. Irretrievable damage can be done to plants and animals by the formation of ice crystals within their cells, probably because of the consequent mechanical damage to membrane structure. Nevertheless, certain fish are remarkable for their capacity to survive beneath the ice of polar waters. This water, at the bottom of salt-water fjords in northern Labrador, for example, is at -1.7 to -1.8°C, which is just above its freezing point, yet fish and invertebrates are able to live there. That invertebrates can do so is understandable, for, as we have seen, they are commonly iso-osmotic with the medium, so that they will not freeze as long as the sea water is above its freezing point. The blood of most teleost fish, however, freezes at about -0.5°C to -0.8°C, and the freezing point in these Labrador fish is only a little lower (-0.9°C to -1.0°C). Why, then, do they not freeze in these cold waters?

Some fish (e.g. the Arctic char, *Salvelinus alpinus*) evade the problem by seasonal migration from shallow Arctic seas to warmer freshwater habitats, but others remain. They are able to do so because their blood is supercooled. This process, which is readily demonstrated in water if it is allowed to cool slowly and without agitation, permits the temperature of the tissues to fall below freezing without the formation of ice crystals. Yet this does not wholly eliminate the danger, for supercooled water rapidly crystallizes on the addition of ice, which is said to 'seed' the formation of the ice crystals. This adaptation, therefore, is only serviceable for fish which are unlikely to encounter ice. If they do come into contact with it, they immediately freeze, as can readily be demonstrated in aquarium experiments. There is thus a behavioural element in this adaptation, which can only operate in animals living well below the ice layer, or which, like the non-polar killifish *Fundulus heteroclitus*, actively avoid contact with the ice which lies at the surface of its habitat in winter.

There are, however, fish which manage to live so close to the ice in arctic waters that they are readily caught by Eskimos for food. These fish exploit another device. Studies of cod (*Gadus*) and sculpin (*Myoxocephalus*) have shown that the freezing point of their

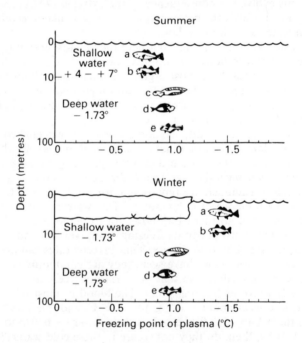

Fig. 7.4 Plasma freezing points of shallow-water fish (a, *Gadus ogac*; b, *Myoxocephalus scorpius*) and benthic fish (c, *Lycodes turneri*; d, *Liparis koefoedi*; e, *Gymnacanthus tricuspis*) in summer and winter. The position of the fishes on the abscissa indicates the freezing point of their plasma. (After Scholander *et al.* (1957), from De Vries, 1971. In *Fish Physiology*, **6**, 157–90, (Hoar and Randall, eds). Academic Press, New York.)

plasma, which is $-0.80°$ in summer, is lowered in winter to $-1.47°$ in the cod and to $-1.50°C$ in the sculpin, presumably in response to seasonal changes in ambient temperature. Other examples seen in Fig. 7.4 show that fish living continuously in deep water, where the temperature does not undergo seasonal change, retain the same low freezing point throughout the year. This adaptation still leaves the fish slightly supercooled, for the freezing point of the water in which they live is $-1.73°C$, but the risk of them freezing is clearly very much reduced. In effect, they have anticipated the device used by man to keep his automobiles useable in winter: the use of antifreeze. The nature of this antifreeze is not always clear, although in some antarctic fish it consists of a group of glycoproteins composed of alanine, threonine, N-acetylgalactosamine and galactose.

Thermal Problems on Land:

Plants

The emergence of life upon land exposed plants and animals to more rigorous temperature regimes than are experienced in water, with heat exchange taking place along the various pathways shown in Fig. 7.1. Plants, as we have noted, can do little to control their temperature, but despite this, and despite their immobility, some can survive extreme cold, while others can flourish even in the harsh environments of hot deserts and semi-deserts, thanks often to adaptations that meet their need to conserve moisture.

Plants adapted to tropical and semi-tropical conditions are apt to be injured by long-term chilling in the range $10–12°C$, the injury probably resulting from changes in the lipid phase of their cell membranes. Plants from colder regions can, of course, survive such temperatures, and may sometimes be highly resistant to freezing. Frost damage, however, is for many plants a serious hazard and a major factor in limiting their distribution.

Dry seeds are usually highly resistant to freezing, but become susceptible when they are hydrated and begin to germinate. Thereafter some degree of acclimatization (p. 131) of the mature plant is often possible, resistance to cold being higher in the winter. Different parts of the plant may also vary in their resistance at any one time, underground parts such as roots and tubers being more resistant than shoots, while dormant buds are more resistant than young leaves.

The basis of frost resistance is obscure, but it is at least clear that in plants, as in animals, the formation of intracellular ice crystals leads to death. A number of factors probably contribute to the

avoidance of intracellular freezing in plants, one being a capacity
for supercooling (p. 135), well seen in winter wheat, which can
supercool to $-25°C$. When ice crystals do form, however, they
commonly first appear outside the cells (extracellular freezing), at
least during slow cooling. This results in water passing out of the
cell to the crystals, and the dehydration may then leave the
protoplasm more susceptible to mechanical injury. The risk of the
movement taking place can be diminished by a rise in intracellular
osmotic pressure, brought about by increase in carbohydrate
content, while changes in the configuration of protein molecules can
confer increased stability against the denaturation that could result
from low temperature and loss of water. Denaturation changes,
leading to damage in the cell membrane, with a consequent spread
of ice formation into the protoplasm, probably account for frost
damage being instantaneous, in contrast to the more protracted
effect of chilling.

Most active plants, like most animals, cannot long survive
temperatures in excess of about $40°C$, for these promote protein
breakdown and a variety of biochemical lesions. Those plants that
can survive high temperatures (thermophiles) are least at risk in
tropical rain forests, for temperatures there are relatively uniform,
the high humidity largely eliminates the risk of excessive
transpiration, while the heating effect of the sun is reduced by the
protective forest canopy. The problems presented by hot deserts,
however, are much more serious.

Plants adapted for average water supplies (mesophytes) can live in
localized regions of hot deserts where the soil retains some
capillary moisture, but the typical plants of deserts are xerophytes
(adapted for prolonged dry conditions); these depend upon a variety
of devices, often based upon maximum exploitation of any water
that may become temporarily available.

One line of adaptation (much as in cold conditions) is to survive
during the hottest season as bulbs, corms or seeds, and to confine
activity to short rainy seasons (which do not necessarily occur every
year), with rapid completion of flowering and seed formation, and
with germination being often promoted by the leaching out of
inhibitors. The dispersal of seeds at such times probably contributes
to the nutrition of animals, which are thus helped to survive even
in the absence of the parent plants to provide a start to their food
chains. Strictly, these xerophytes are evading drought, and so also
are perennials such as the acacia, with long tap roots that penetrate
many feet underground to reach deep-lying water tables (Fig. 7.5).

Many plants can absorb moisture from the night air after its
condensation on the stems and leaves. We shall see that such
absorbed water is important also in the water economy of some

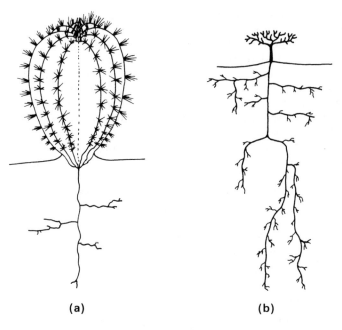

(a)   (b)

Fig. 7.5 Rooting system (**a**) of a cactus and (**b**) of an acacia, from the side. Not to scale.

desert animals, for which it may be the only source of moisture, apart from their own water of metabolism. In such ways the lives of animals and plants run on parallel or intersecting lines, as they also do, although in different patterns, during brief flowering periods, which are correlated with the presence of pollinating insects. Another device (found in cacti, for example) is the development of water-storing tissues, which make it possible to extend the use of the water falling during the rainy season. Such plants, called succulents, have shallow and wide-spreading rooting systems (Fig. 7.5) which are well adapted to benefit from light rain that may not have penetrated far into the soil. Succulents further conserve water by having a low rate of transpiration, dependent on a thick cuticle and on their stomata being open at night and closed during the day. This is associated with metabolic adaptations that permit the synthesis of organic acids by night and their decarboxylation during the day, so that carbon dioxide is made available for daytime photosynthesis.

Some perennial xerophytes are so drought-resistant that they can survive prolonged periods of wilting and desiccation; these are often regarded as the only true xerophytes. Their adaptations include a limited size of shoot and a disproportionately large root area, an ability to take up water from soil that is relatively dry, a thick cuticle and sunken stomata. During wilting, the leaves of some species become vertically orientated, with consequent reduction of the amount of solar radiation that they receive.

The heating effect of solar radiation can be moderated, of course, by the evaporative effect of transpiration, and this is sometimes of critical importance in the Sahara desert. As with amphibians (p. 144), however, the scope is limited by the need to conserve water. Xerophytes are helped in this respect by being, in general, better able than mesophytes to withstand dehydration, a capacity which is expressed as the critical water saturation deficit. This is the water loss, expressed as a percentage of the water content at full saturation, that results in death. For true mesophytes the value is below 25; for true xerophytes it is greater than 75. In practice, however, the need to conserve water predominates, so that true xerophytes tend to have, in addition, a low rate of cuticular transpiration, at least during dry periods, although it may become high during wet ones.

Invertebrates

The organization and the mobility of animals has enabled them to evolve a variety of responses to the rigour of terrestrial conditions. We can consider here only a few illustrations of the diversity and flexibility of these, contrasting particularly those of poikilotherms with those of homeotherms, and concentrating mainly upon life under extreme conditions, which brings out so clearly the adaptive flexibility of animals. We shall see that life under variable and often rigorous regimes demands the use of metabolic processes as well as the exploitation of the physical factors (conduction, convection, evaporation and radiation) that we have already outlined. With these are associated behavioural reactions, by which animals are enabled to develop this exploitation to their maximum advantage.

Commonly they avoid the extremes of solar radiation by burrowing or seeking other forms of shelter, but there are exceptions to this, one remarkable one being provided by a desert snail of the Near East, *Sphincterochila boisseri*, which seems positively to welcome full exposure (Fig. 7.6). This snail, which is found on finely particulate loess, comes out after rain and feeds by ingesting the substratum, but it is also often found on the substratum in a dormant state, fully exposed to the sun, when the surface temperature of the loess may reach a maximum of 65°C. Several factors make possible this exposure. The maximum temperature reached by the loess underneath the snail is only 60°, because it is shaded by the snail's shell. Further, the animal's body

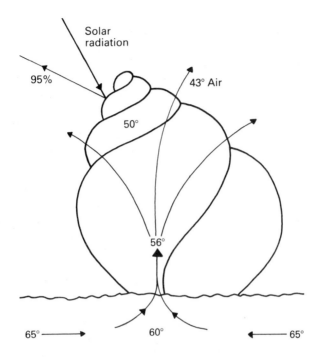

Fig. 7.6 The temperature in a dormant snail (*Sphincterochila*) on 9th July, 1969, at the highest temperature recorded during July and early August. (From Schmidt–Neilsen *et al.* (1972). *Symp. zool. Soc. London*, **31**.)

is withdrawn into the smaller (upper) whorls of the shell, where the maximum temperature reaches only 50°C, because the air pocket in the larger (lower) whorl insulates it from the loess surface. The value of this insulation can be shown by filling the whorl with water; the result is to raise the temperature of the body by some 5°C, which, since the lethal temperature for the snail is between 50° and 55°C, could well prove fatal. Like the locust mentioned later, the snail is exploiting the low thermal conductivity of air, the two animals thus presenting an instructive illustration of how the same physical property can be exploited in two quite different patterns of organization. The locust climbs up a plant; the snail climbs up the inside of its shell. Even so, these factors do not wholly account for the snail being so much cooler than the substratum. A further one is the high reflectance of the shell surface, which dissipates over 90 % of the incident radiant energy.

Physiological adaptations also contribute to the survival of the snail when it is in a dormant state. The metabolic rate is very low, so that the stores of carbohydrate and fat, although small, are sufficient to provide for survival over several years, supposing that this proved necessary. As for water content, this amounts to 1400 mg, and, since the observed rate of water loss is only 0.5 mg per day, this could again ensure survival for several years.

Insects, with their very high capacity for movement, present a different pattern of adaptations. These animals, which have shown outstanding versatility in every aspect of terrestrial exploitation, are, of course, ectothermic, and are thus very much dependent upon the absorption of solar energy, although their own generation of heat can become very high during flight. At low temperatures they cannot fly, but, like vertebrate poikilotherms (p. 145), they can warm themselves by uptake of solar energy, as exemplified by butterflies exposing their wings at right angles to the path of the sun's rays. They are thus using a behavioural response to exploit a physical process, a type of adaptation illustrated in more precise terms by the thermal responses of the desert locust (*Schistocerca*).

Inactive at low temperatures, this insect begins, at about 17°C, to align itself at right angles to the sun's rays and, by leaning over, to expose as much of its body as possible to them. At temperatures of about 40°C it is in danger of overheating (its lethal temperature is at about 45°C), and it now orientates itself parallel to the sun's rays, thereby reducing the incidence of solar radiation. It can further react by stretching its legs to raise its body above the ground (Fig. 7.7), or by climbing up vegetation. The effect of this is to reduce the uptake of heat by conduction from the hot substratum, while it is also exploiting the very low thermal conductivity of air (0.000 05 as compared with 0.0014 for water), which results in a rapid fall in temperature above the desert surface.

Insects, like other terrestrial animals, can also lose heat by the evaporation of water from their spiracles as a result of respiration,

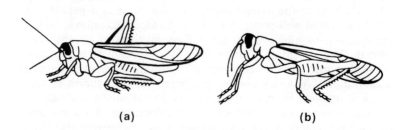

(a)   (b)

Fig. 7.7 Temperature control in locusts. The insect to the left, in the early morning, is leaning over, broadside on to the sun. The insect to the right, in the middle of the day, faces the sun and raises the front of the body.

but this process is limited in value, at least in dry conditions, by the consequent reduction in body water, which must lead eventually to death from desiccation. Water loss from the insect body is restricted by the waterproofing of the cuticle, which, in desert forms, is almost impermeable, but at high temperatures there is a change in the condition of the waterproofing layer, resulting in an increased rate of water loss (Fig. 7.8). It follows that unless the need for cooling the body is imperative (and insects can, in fact, lower the temperature of their body by as much as 5°C below that of their surroundings when the need arises), these animals must limit the opening of their spiracles, and rely upon behavioural adaptation, including the selection of shelter, to guard them from excessive water loss.

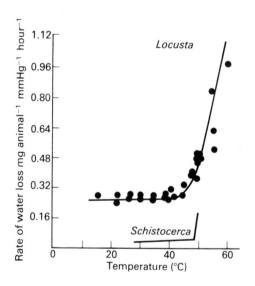

Fig. 7.8 The relationship between cuticular transpiration and temperature in *Locusta* and *Schistocerca*. (From Shaw and Stobbart (1971). *Symp. zool. Soc. London*, **31**.)

One remarkable example of the exploitation of thermal loss by evaporation, although not from the insect's own body, is found in hive bees, which, in a combination of physical and behavioural thermoregulation, cool their hives by transporting water into them and then evaporating it by fanning it with their wings. In this way the temperature of the hive can be lowered to 36°C when the outside temperature has risen to 40°C. The ingenious organization of their social life also provides for raising the temperature of the hive in cold weather. This they do by clustering together and generating heat by movement within the mass.

Flight in insects imposes a special thermal problem, for it can lead to a 50-fold increase in heat production. Some of this heat is lost by evaporation and radiation, but most by convection, as might be expected when the body is moving so actively. There are, however, circumstances when it is disadvantageous to lose this heat too rapidly. This is so with moths, which, flying at night, need to warm their bodies to a temperature suitable for activity. Many of them achieve this through the development of hairs on their thorax, which maintain a stable air layer around the body surface and thus reduce heat loss by convection; an instructive analogy with the body covering of birds and mammals.

Amphibians and reptiles

The thermoregulation of terrestrial vertebrates, like their mechanisms for ion and water regulation, exemplify stages in progressive adaptation to the rigours of terrestrial life, culminating in the fundamentally important transition from the poikilothermy of amphibians and reptiles to the homeothermy of birds and mammals. Of the poikilotherms, the amphibians are the more restricted because of their permeable skin. Those that are permanently aquatic approximate closely to the ambient temperature, and react much as fish, selecting suitable regions, and being capable of temperature compensation. Amphibians on land lose water by evaporation through their skin so freely that they behave as wet-bulb thermometers when they are in air that is moving and unsaturated. Like insects, they can take up heat from solar radiation, but this is only practicable in environments which permit replacement of the consequent water loss. Their risk of desiccation is, however, partially compensated for by a complex of responses termed the water-balance or Brunn effect. This comprises a reduction of urine flow, uptake of water from the bladder, and absorption through the skin, all promoted and coordinated by a hormone, arginine vasotocin, which corresponds to the vasopressin of mammals. Significantly, the effect is best developed in species inhabiting drier environments. The exploitation of this adaptation enables amphibians to survive even in hot deserts; here their survival is further aided by retreat to underground burrows, where temperatures are much lower than on the surface, and by a low metabolic rate, which permits them to remain inactive for several years during prolonged drought. One may marvel as much at the competitive pressures which drive animals into such extreme conditions, as at the fertility of natural selection in solving the resultant environmental problems. A particularly remarkable example of this fertility of invention is a Rhodesian frog (*Chiromantis*) which is aided in its tolerance of an arid environment by the rate of evaporation of water from its skin being reduced to a very low level. This rare exception to the normal pattern of amphibian organization is, as we shall see, much more characteristic of reptiles.

These animals differ fundamentally from amphibians in thermal relationships because, with the exception of a few secondarily aquatic forms such as crocodiles, they have lost the buffering effect of an aquatic environment which is still a factor in the lives of most amphibians. They have to deal, therefore, with the full rigours of terrestrial temperatures. This they do in ways which foreshadow the homeothermic devices of birds and mammals, and which were probably carried much further in extinct forms. It is likely that some of the giant dinosaurs were homeothermic, for their large body size, with relatively small surface area, would have favoured the retention of heat. Pterosaurs (flying reptiles) were perhaps also homeotherms; some are said to have had a hair-like covering, and it is to be expected that elevated body temperature would have facilitated their flight by promoting a high rate of metabolism.

We have seen that improvement in the respiratory mechanism enables reptiles to have a largely impermeable skin, but it is not completely impermeable; evaporation can therefore take place through it to an extent which is correlated with the normal environment. Thus crocodiles lose water by evaporation at a rate up to as much as one-half of that in amphibians, while the rate in a desert lizard, *Sauromalus*, is only 5 % of the crocodile's. Much of this evaporative loss (two-thirds or more) takes place through the skin, but the respiratory system also contributes. This makes possible some regulation of water loss, giving reptiles an advantage over amphibians, which cannot control their evaporative water loss through the skin. The respiratory rate increases at high temperatures, powerful pumping movements of the mouth and neck developing in the large monitor lizard *Varanus*. This is the functional equivalent of panting in mammals, providing for an increased rate of cooling of the body. *Varanus* can also raise its body temperature by increasing its metabolic rate, to an extent which is said to exceed the basal heat production of a mammal of equivalent size. These responses show that reptiles use metabolic and physical processes for thermoregulation in ways that anticipate the homeothermal organization of mammals. The resemblance is further increased by circulatory adjustments, which permit reptiles to conserve heat or to distribute it, as needs dictate. We shall see later the nature and importance of such adjustments in mammals.

However, behavioural responses are of prime importance in

reptiles. It has been best studied in desert-living species, where it enables them to maintain a surprisingly high level and constancy of body temperature which belies their description as poikilothermal. Each species has a characteristic preferred temperature, and maintains this for a considerable part of the day by means of behavioural responses to heat fluxes in the surroundings. These responses are probably integrated with internally regulated (endogenous) diurnal rhythms, ensuring that the animal is awake and prepared to take advantage of the ambient temperature, and it is possible that seasonal rhythms are also involved, for there is evidence of seasonal changes in preferred temperatures.

A well-studied example of such responses, strikingly reminiscent of those of the desert locust, is the earless lizard of the South Western United States. By night it is concealed in its burrow in the sand, but in the morning it thrusts its head outside. The blood flowing through a large sinus in the head now takes up heat which is distributed through the body so that this becomes warmed to a temperature appropriate for full activity. The lizard then emerges, and thereafter keeps its temperature approximately constant by sheltering, or by resting on the substratum, as may be appropriate. Here again there is a foreshadowing of the homeothermy of birds and mammals, although it is, of course, attained by quite different means. The lizard is evading, by behavioural means, those fluctuations of body temperature which would otherwise be imposed by its environment, and is thereby securing the advantage of a high and comparatively constant level of activity during its waking hours. This advantage is gained, in desert conditions, at low metabolic cost, for the animal has little to do beyond exploiting solar radiation. In other conditions (in tropical forests, for example) the cost could be higher, but the advantage would still be great enough to justify this. Indeed, there must have been very strong selection pressure during the emergence of birds and mammals to secure this advantage by developing a high basal metabolic rate, in conjunction with the development of feathers, hair and sweat glands, which were to be the crucial new developments.

Birds and mammals

Birds and mammals evolved from two distinct reptilian stocks along independent lines, but their homeothermal mechanisms are sufficiently similar to justify treating them together. Both groups maintain a constant and high core temperature by achieving a balance between the uptake of heat from solar radiation, the production of heat within the body, and the loss of heat from the body to the environment. That their core temperature is a high one is the consequence of two factors. One is their endothermy, made possible by their high basal metabolic rate, a parameter which is expressed as the oxygen consumption of the resting animal per unit weight per unit time. As an illustration of this, the rates for an ectothermic rattlesnake weighing 2.5 kg is 32.2 kJ kg^{-1} 24 h^{-1}, while that of an endothermic rabbit of the same weight is 187 kJ. The second factor is the insulation of the body surface, provided by feathers in birds and hair in mammals (clothing taking its place in man), supplemented by fat deposits. (It may well be that feathers first evolved, like hairs, as a thermoregulatory device, and only later became a crucial factor in the evolution of flight.) This insulation tends to the conservation of heat within the body, and contrasts markedly with the arrangement in poikilotherms, which typically have their fat deposits deeper in the body, where they do not interfere with the uptake of solar radiation through the skin. The constancy of core temperature that also distinguishes birds and mammals from reptiles is maintained by involuntary nervous controls, which means, as far as we are concerned, that they are not consciously directed and are independent of our behaviour. As in reptiles, however, behavioural responses do have an important thermoregulatory role.

The production of heat is increased by exercise, but this is not primarily a thermoregulatory process; rather it evokes thermoregulatory responses tending to promote heat loss, but it does become important in cold conditions when additional heat production is required to maintain the normal core temperature. In addition to voluntary exercise (activities, that is, that are under the control of the will), cold conditions also invoke the involuntary muscular tremors called shivering, which can produce a five-fold increase in heat production. This can also be increased in the cold by metabolic (non-shivering) thermogenesis, evoked by certain hormones which exert what is called a calorigenic action. These hormones are adrenaline, secreted by the medulla of the adrenal gland, and producing a relatively short-lived effect, and the hormones of the thyroid gland (thyroxine and triiodothyronine), which produce a more prolonged action. Finally, the comforting effect of a warm meal in cold conditions is due in part to a calorigenic action of the food, termed its specific dynamic action, which results in an increased metabolic rate following the absorption of the digestion products into the blood stream. The effect is evoked particularly by the protein component of the food, but carbohydrates and fats also contribute.

The regulation of heat loss depends upon modifying the temperature gradient between the core and the body surface, and between the latter and the environment, by exploiting and manipulating the physical processes of conduction, convection, radiation and the evaporation of water. Because of the high core temperature of homeotherms, there is normally a flow of heat from the body to the environment, and it is here that feathers and hair are important. They trap at the body surface a layer of air which, because of its low thermal conductivity, reduces the loss of heat from the skin. Moreover, heat retention can be increased in the cold by erecting the hairs and feathers, which increases the thickness of the trapped layer. The erection is effected in mammals by arrector pili muscles inserted on the hair follicles; it is the contraction of these muscles which causes 'goose-flesh' in man.

Heat loss from the body surface can be further regulated by controlling the flow of blood to the skin. In cold conditions, the most superficial vessels are constricted and blood flow is largely confined to the deeper vessels. The conservation of heat is further promoted because venous pathways run close to the superficial arteries, making possible a counter-current exchange system, similar in principle to that of fish respiration (p. 98). Heat passes from the arterial blood to the venous, so that the former is precooled before it reaches the body surface, while the venous flow takes back warmed blood to the interior of the body.

Conduction is often of little importance, although it can lead to considerable loss of heat from the surface when the body is in contact with a cold substratum, or when it is immersed in water. The potentially disastrous effect of the immersion results from the displacement of the insulating layer of air by the water, with its higher thermal conductivity; which is why the polar bear shakes itself so vigorously when it leaves the water. However, heat loss by immersion is probably one of the benefits gained by wallowing in mud or water (seen, for example, in the hippopotamus, and appreciated also by pigs, which, lacking sweat glands, rely largely upon it). No doubt the cooling produced by evaporation of the water from the body surface on emergence is also beneficial to these animals, unwelcome though it may be to us after swimming.

Heat loss by convection is promoted by the low specific heat of air, which is rapidly warmed by contact with the body surface. The warmed air rises and is replaced by cooler and denser air, so that the temperature gradient continues to favour heat loss. The process, as we well know from our own experience, is encouraged by wind and draughts. Loss of heat by radiation to solid bodies in the neighbourhood depends upon differences in the temperature of those bodies and the skin. The loss may be considerable, amounting in man to as much as 50 % of the total heat loss during comfortable resting.

Loss of heat by evaporation occurs in two ways. One is the

insensible water loss, due to the loss of water vapour through the skin and in the expired air from the lungs. The other is sensible or regulated water loss, which takes place by panting and, in many mammals, by the release of sweat from the sweat glands of the skin. Sweating, which in man begins at ambient temperatures of about 29°C, depends for its effectiveness on the humidity of the air. In warm humid conditions the sweat cannot evaporate and it therefore accumulates on the skin as a fluid which contributes nothing at all to heat loss. In these conditions panting becomes important, as it does also in mammals which have few sweat glands, and in birds, which have none. Panting, which is essentially a high rate of shallow breathing, promotes the movement of air over moist respiratory surfaces, and over the lining of the buccal cavity. The evaporation of water is thus promoted, and is further aided by increased salivation.

Thermoneutral Zones and Critical Temperatures

The complex interaction of the processes which have been outlined can be illustrated by a generalized statement of the responses evoked when a mammal is subjected to a range of ambient temperatures. There will be a zone of temperature, varying in range from species to species, which permits the animal to maintain its lowest rate of metabolism. This zone, called the thermoneutral zone, is characterized by the body temperature being maintained constant by physical processes alone, in conjunction with the regulation of skin temperature. When the ambient temperature rises above this zone, there are two possibilities. Behavioural responses may take the animal to a cooler shelter. If, however, they do not, then the body temperature will rise (hyperthermia) and heat death must eventually result. When the ambient temperature falls below the thermoneutral zone, body temperature must be maintained by increasing metabolic rate, the temperature at which this increase begins being termed the critical temperature. The process is metabolically expensive, and must lead to exhaustion if the increased use of energy reserves cannot be made good by increased intake of food. This is why critical temperatures vary markedly from species to species, and show close adaptation to the normal environment. Tropical mammals like the night monkey (Fig. 7.9), and also tropical birds, tend to have narrow thermoneutral zones and high critical temperatures; they are said to be temperature sensitive, and can afford to be because they do not normally experience low temperatures. A naked man falls into this category, which suggests that we are tropical in origin. It has been well said that we carry our

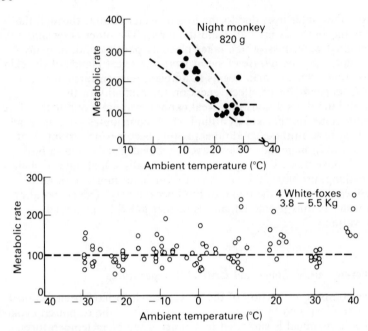

Fig. 7.9 Heat regulation and temperature sensitivity in arctic and tropical mammals. The fox needs only a slight increase in metabolic rate to withstand the coldest temperature on earth. Basal metabolic rate = 100. (After Scholander *et al.* (1950). *Biol Bull.*, **99.**)

tropical environment with us in the form of clothes and heat technology.

Large arctic mammals, such as the arctic fox and the arctic dog, provide the greatest possible contrast (Fig. 7.9). With broad thermoneutral zones, and critical temperatures of around − 40°C, a small increase in metabolic rate serves to maintain their body temperature under conditions of extreme cold. So it is that the arctic dog, with good insulation provided by a thick coat of hair, an efficient regulation of skin temperature, and behavioural responses promoting suitable postures, can actually sleep on open snow at an ambient temperature of − 40° C. There is, however, a price to be paid. If such animals are transported into a warm environment, they find it difficult to dissipate heat, and may even die of heat stroke.

Small homeotherms cannot emulate the larger ones in these responses to extreme cold, for a small body size gives a large surface area relative to volume. This encourages heat loss, and at the same time limits insulation by precluding long hair and a thick insulating coat. Metabolic thermoregulation would be an impossibly extravagant response, but there are other paths of escape. Birds can resort to migration, while small mammals can find warmer temperatures by burrowing into the snow (so, too, does the ptarmigan, *Lagopus*). But perhaps the most effective way to conserve energy in the cold, not only in polar regions but in temperate ones as well, is either to sleep or to hibernate.

Animals resorting to deep sleep, such as the brown bear, remain fully homeothermal, but hibernation (seen, for example, in the hedgehog, the marmot and the hamster) is altogether different in that it depends upon far-reaching physiological adaptations. The hibernating hamster undergoes a marked drop in body temperature and a decline in basal metabolic rate to levels as low as 1 % of normal. The body temperature of a hibernating mammal varies with the environment, but it is not poikilothermic, for it can raise its metabolic rate to avoid freezing when the ambient temperature falls dangerously low. There is adaptive specialization here which is lacking in reptiles, for these cannot raise their metabolism, and can only survive in a freezing spell if they have selected a sufficiently protected site for their overwintering. A further complex of specializations comes into operation when hibernating homeotherms awaken. There is a rapid rise in body temperature, brought about by shivering, which is quite independent of the ambient temperature. Warming of the heart promotes the circulation of the blood, which carries heat with it, and is restricted in such a way that the heat passes largely to the respiratory system and the brain during the early stages of wakening. This is important, particularly for the brain, which, being highly sensitive to temperature, could be irrevocably damaged if it was chilled as it came into full function. Heat is also generated at this time by the metabolism of deposits of brown fat, the cells of which differ from those of the more usual white fat in having many mitochondria associated with their numerous fat droplets.

Mammals in Hot Climates

Consideration of homeotherms in the cold has brought us close to one of the limits of their survival. No less exacting are the problems presented to them by extreme heat. Extreme cold can be met, as we have seen, by conserving heat, but, given that the core temperatures of homeotherms lie within the range of 35–42°C, consider the implications of a mean maximum temperature in Baghdad of 43.6° and of a maximum shade temperature of 58°C recorded at El Azizia, Libya. Clearly in hot conditions there may be a flow of heat from the environment into the body. This heat can be dissipated in arid conditions by evaporation, but survival will then depend on replacing water loss, so that water needs to be conserved in every way possible. In hot humid conditions, by contrast, the scope for evaporation is restricted, with consequential heat stress and discomfort familiar to all who have had to endure such climates. Large size in mammals can help, however, for uptake of heat is then minimized because the surface area is small relative to volume. Moreover, the heat inertia which goes with large size creates the possibility of temporary heat storage without undue rise in temperature, while larger animals are better able than smaller ones to search for water holes.

In general, the physiological consequences for mammals of life in hot conditions have been much more fully studied in arid climates than in humid ones, and for this reason reference will here be made only to life in hot deserts, where, as we have already seen, poikilotherms, invertebrate and vertebrate, and plants as well, have found means of survival. Two well studied examples, strongly contrasted, will serve to indicate the flexibility of homeothermal adaptations to desert life, and the important contribution made thereto by behaviour patterns.

Amongst the more mobile members of the desert fauna, the camel is doubtless the most famous, with an importance to man that needs no emphasis. Contrary to popular belief, it does not store water, so that we must forego the otherwise appealing analogy with succulent cacti. Instead, it simply drinks it to compensate for dehydration, doing this with a speed and precision that cannot be matched by man. During the intervals between drinking, its survival depends upon regulation of its water loss. Under experimental conditions, camels can survive the loss of 20–22% in body weight by dehydration. Their tissues must therefore have remarkable powers of resisting dehydration, but the cellular mechanisms involved are not understood. Under normal conditons it relies on reducing the loss of water in its faeces by absorbing it from the alimentary tract, while it reduces excretory loss by producing a highly concentrated urine, at 3000 mosmole kg^{-1}. Its efficiency is emphasized by contrasting it with the Somali donkey (*Equus asinus*), which is herded with the one-humped camel (*Camelus dromedarius*) by nomads in northern Kenya. The donkey has a significantly lower absorption of water, and produces a less concentrated urine (1500 mosmole kg^{-1}). One result is that the camel can thrive when drinking salt solutions more concentrated than sea water (up to 5.5% NaCl), whereas the donkey cannot exist on these.

Perhaps the most interesting adaptation of the camel, however, is its capacity, during dehydration, to tolerate wide variations in its core temperature, which may range from 41°C by day to 36°C by night. It was supposed at one time that this was a primitive form of heterothermy, but it is now recognized as a valuable environmental

Table 7.1 Maximal urine concentrations in various terrestrial mammals. The figures are the highest available, but do not necessarily represent the concentration limits for the species. Electrolyte concentration is an approximation referring, in some cases, to Na concentration, in others to NaCl equivalence from electrical conductivity. (Taken from Gordon, 1979. Table 7.12, p. 279. Source is Table XXVI in Schmidt-Nielsen. *Desert Animals*. Oxford University Press.)

Animal	Urine Osm 1^{-1}	Urine electrolyte meq 1^{-1}	Urine : plasma osmotic ratio
Homo sapiens (man)	1.43	460	4.2
Rattus rattus (white rat)	2.9	760	8.9
Camelus dromedarius (camel)	2.8	1070	8.0
Citellus leucurus (antelope ground squirrel)	3.9	1200	9.5
Gerbillus gerbillus (gerbil)	5.5 (est.).	1600	14.0 (est.)
Psammonys obesus (sand rat)	6.34	1920	17.0

adaptation, resulting in a reduction of evaporative water loss. The principle is somewhat analogous to heating a house with storage heaters. During the day the camel stores heat and thereby saves evaporative water loss which would otherwise be expended in cooling the body; it has been calculated that a 500 kg camel could thus save 4 l of water. During the night it loses heat, but again conserves water by making use mainly of conduction and radiation.

A contrast to the camel is provided by the kangaroo rat, *Dipodomys merriami*, representative of the small burrowing rodents which survive in the desert largely by behavioural adaptations, associated with rigorous control of water relations (Fig. 7.10). During the day it lives in burrows where, at a depth of 1 m, the temperature may be only about 30°C when the surface temperature is reaching 80°C. It has been calculated that if it did not do this, its evaporative water loss would rapidly prove fatal, for it would amount to about 10% of its body weight per hour. Like the camel, it regulates its water turnover, but along different lines, being able to maintain itself in water balance without any drinking at all, provided that there is a small quantity of moisture in the atmosphere. (We have earlier seen xerophytes exploiting the same principle.) With this condition satisfied, there will be sufficient moisture on the vegetable food which it obtains at night, to compensate, in conjunction with the water produced by its own metabolism, for water loss in the faeces and urine, and by evaporation. This is possible because the loss is small, contributory factors here being a hyperosmotic urine and a

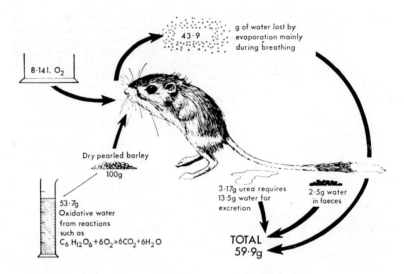

Fig. 7.10 The water relations of *Dipodomys*, the kangaroo rat, when fed on dry pearled barley. The difference of 6.2 g between the total water loss of 59.9 g and the water derived from oxidation of food is made good if the pearl barley is in equilibrium with air of 20 % relative humidity. (Based on Schmidt–Neilsen (1964). *Desert Animals*. Clarendon Press, Oxford; from Chapman, 1967.)

counter-current heat exchange system in the respiratory passage. The temperature of the exhaled air is reduced by this below that of the inhaled air, so that water condenses in the exhaled air and can be reabsorbed. A further contributory factor is the relative humidity of the burrow, which is higher than that of the outside air, a factor which we have seen to be important in the lives of desert-living amphibians. The value of the sum total of these adaptations is seen when *Dipodomys* is compared with the laboratory white rat. Water loss is substantially higher in this animal, which would be rapidly dehydrated if it were placed in the same conditions as those in which the kangaroo rat readily survives.

The significance of the composition of the urine is readily apparent. Man and the white rat, in which it is relatively dilute, require frequent access to drinking water. The gerbil (of Old World deserts), producing a highly concentrated urine, can survive on air-dried food. The sand rat (Saharan desert), prone, like the gerbil, to evaporative water loss, produces an even more concentrated urine. It, however, obtains its water from succulent plants with a high salt content, but a low nutrient one. Large quantities of these plants must be eaten because of this, and correspondingly large amounts of salt eliminated in the urine.

8 Life Histories

Evolution and Sex

We emphasized in Chapter 1 the effectiveness of mitosis in ensuring precise replication of the genome. It is now necessary to examine how modifications of the genome can bring about that variation in the characters of the organism which is essential for the improvement of adaptation by natural selection. To do this, we must examine some aspects of reproductive processes, and the ways in which these are deployed in plant and animal life histories. Here we have to distinguish two modes of reproduction, asexual and sexual, which differ fundamentally in their impact upon hereditary variation.

In asexual reproduction, found commonly in plants, but much less so in animals, parts of the body, or individual cells, separate from the parent and develop directly into new organisms. They are thus essentially extensions of the parent, and have exactly the same genetic constitution as that parent, except in the rare event of a mutation (p. 19) having taken place. All of the descendants of a single individual which have been produced by uninterrupted asexual reproduction, and which therefore share this common genetic constitution, are termed a clone.

Sexual reproduction, found in the vast majority of plants and animals, depends upon a complex sequence of events which is essentially the same in both kingdoms. This similarity in detail makes it likely that sexual reproduction evolved in the common ancestors of plants and animals, at a very early stage of the history of life. The reason for its early origin and subsequent almost universal persistence lies in the vastly increased scope which it provides for the promotion of hereditary variation, and hence for the evolution of new or improved characteristics through the action of natural selection. Full details will be found in textbooks of genetics, but the principle involved may be stated in simple terms.

Sexual reproduction typically involves the production by two parents of reproductive cells (gametes). Each gamete contains a single set of chromosomes (haploid condition). The gametes fuse in pairs, one gamete from each parent, to form cells called zygotes. These contain two sets of chromosomes (diploid condition), each chromosome in one set being matched by a partner (homologous chromosome) in the other set (cf. Chapter 1). There is much diversity in the relation between the diploid and haploid phases, and, in particular, a fundamental difference in this respect between the life cycles typical of animals and those typical of plants.

In animals it is usual for the diploid zygote to produce, by cell division, a diploid adult phase which then produces gametes:

$$\text{Diploid zygotes} \rightarrow \text{Diploid adults} \rightarrow \text{Haploid gametes}$$

During the divisions that produce these gametes, there is a form of cell division (meiosis) in which the haploid condition arises (Fig. 8.1). This is achieved by the pairs of undivided chromosomes coming together (synapsis) and subsequently separating (segregation), so that two haploid daughter cells are produced. Commonly this division is quickly followed by another in which the haploid set divides by

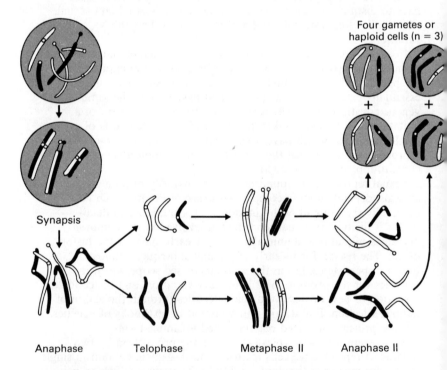

Four gametes or
haploid cells (n = 3)

Synapsis

Anaphase   Telophase   Metaphase II   Anaphase II

Fig. 8.1 Diagram of meiosis in an organism having three pairs of homologous chromosomes (2n = 6). (From Alston (1967). *Cellular Continuity and Development.* Scott, Foresman, Glenview, Illinois.)

mitosis so that four haploid gametes have arisen from the initial
mother cell. This suggests that the process of chromosome reduction
has evolved by the modification of two successive cell divisions so
that the chromosomes have come to divide only once while the cell
divides twice.

In plants the diploid zygote usually produces a diploid sporophyte,
which reproduces asexually by producing haploid spores. These are
called meiospores, because their production has involved meiosis.
The spores give rise to haploid gametophytes; these produce haploid
gametes which complete the life cycle by fusing to yield diploid
zygotes:

$$\text{Diploid zygotes} \rightarrow \text{Diploid sporophytes} \rightarrow \text{Haploid spores} \rightarrow \text{Haploid gametophytes} \rightarrow \text{Haploid gametes}$$

Here, then, there are two contrasting reproductive generations (one
or other of which may be quite inconspicuous); a sequence of events
termed alternation of generations. This type of life cycle has been
subject to much evolutionary modification, largely in adaptation to
the exploitation by plants of terrestrial life, and it is from this point
of view that we shall later consider it. We shall then be able to
appreciate how two contrasted patterns of life cycle have enabled
animals and plants to exploit to the full, and yet in different ways,
the possibilities of aquatic and terrestrial life.

The contribution of the sexual process to the production of
variation results from it being a matter of chance which member of
any one pair of chromosomes goes into which gamete. This is
termed independent assortment of the chromosomes (Fig. 8.1). The
haploid set of any one gamete is thus likely to have a combination
of chromosomes different from either of the two sets of the parent.
This recombination, resulting from segregation and independent
assortment, in conjunction with the fact that the fusion of a gamete
from one parent is a random process, leads to certain characteristics
of the offspring differing in a random way from those of the parents,
although, of course, the basic features of the species persist. (The
reason for this is, essentially, that the genes determining such
features are likely to be identical in both parents, whereas variant
characters will be determined by the allelic differences discussed in
Chapter 1.) Examples will illustrate the consequence of this.

If the parents have only one pair of chromosomes, their
segregation will produce two types of gamete; random fusion of a
gamete from one parent with one or other of the two types
produced by the other parent results in four possible combinations
of chromosomes in the zygote. If the parents have five pairs of
chromosomes, segregation and independent assortment can produce
32 different types of gamete, giving 1024 possible combinations. If

there are ten pairs of chromosomes, the values become respectively 1024 and 1 048 576, and in the general case, with x pairs of chromosomes, the values are 2^x and (2^x).[2] Since humans carry 23 pairs of chromosomes, it is not surprising that offspring (including non-identical twins) differ from their parents. Identical twins, however, which are formed by the division of a single zygote, will have the same genetic constitution.

The above values are minimal ones, for each chromosome carries a number of genes, and the occurrence of allelic differences between pairs of genes greatly increases the variability resulting from the segregation and independent assortment of the chromosomes. Moreover, there may be exchange of genes between the paired chromosomes during meiosis (crossing over), although this is to some extent counter-balanced by the tendency of genes on one chromosome to remain together (linkage). In extreme cases, as we saw in Chapter 1, a group of genes may remain completely linked together as a supergene.

Asexual Reproduction

Asexual reproduction is seen at its simplest in bacteria, unicellular algae, and protozoans, taking place by fission. The cell divides into two daughter cells, which separate and grow until each resembles the parent, now represented only by its offspring. Budding, as in yeasts, is a variant of this, in which the parent retains its own identity while producing offspring which may remain temporarily attached to it to form a chain.

Asexual reproduction in a more advanced form is common in the fungi, being often more important than the sexual method, presumably because it provides for very rapid multiplication. The asexual reproductive bodies are spores, commonly unicellular, produced by the hyphae in enormous numbers, especially at seasons when climate and nutritive resources are favourable. In the common bread mould, *Rhizopus* (Fig. 8.2), a form of sexual reproduction is effected by the fusion of parts of hyphae, which must be + and − (see p. 161). Mingling of the contents and fusion of opposite nuclei leads to the formation of a resting stage, the zygospore. This eventually germinates, with meiosis, to form stalked sporangiophores bearing swollen sporangia, from which large numbers of sporangiospores are eventually released, to be distributed by air currents. In some other fungi (e.g. *Penicillium*) the spores are formed in chains called conidia. Distribution of the spores, in these and many similar cases, is through the air, but aquatic fungi, like aquatic algae, can produce motile spores (zoospores) which swim by means of flagella.

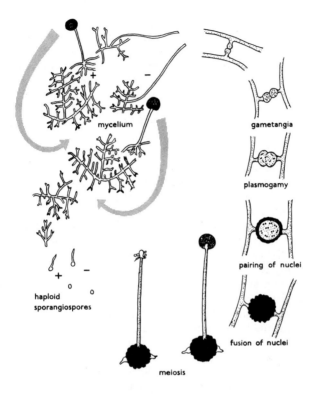

Fig. 8.2 Diagram of the life cycle of the fungus *Rhizopus*. (From Stevenson, 1970.)

Another form of asexual reproduction, and one particularly characteristic of plants, is vegetative reproduction, in which a substantial part of the parent body is detached. This process (which is extensively exploited for cultivation by man) occurs in algae, fungi and non-vascular plants, but is seen at its most varied in the vascular ones, where it may be effected by roots, stems, leaves, and sometimes even by flowers, although these last are primarily organs of sexual reproduction. It may involve the appearance of structures such as roots in unusual positions, in which case they are said to be adventitious. The rapid spread of many grasses, and the part that they play in the consolidation of sand dunes, for example, is aided by their creeping underground stems (rhizomes), from which aerial shoots and adventitious roots are formed. Death of the intervening parts, or their severance by cultivation, results in the separation of new individual plants, to the confusion of careless gardeners. Stems creeping above ground, as in the strawberry, function in the same way.

Very similar in principle is the production of long drooping stems by raspberries, currants and wild roses, to name only a few examples; if these touch the ground, adventitious roots can form and the adjacent buds then grow into upright shoots. Thickets formed in this way are only too familiar. Other examples of the participation of stems are the tubers of the potato, which are the swollen ends of rhizomes, the bulbs of onions, which are large buds with fleshy leaf bases, and the corms of the crocus, which are similar in principle but with small scale leaves. Horizontal roots of many plants behave much as do rhizomes, except that they form adventitious buds instead of adventitious roots. Examples of this are the shoots (suckers) that form from the roots of trees, encouraging the spread and regeneration of timber. English elms have developed extensively from such suckers. They therefore have a high degree of genetic uniformity, resulting from their restricted parentage, and this accounts for their widespread susceptibility to the new strain of Dutch elm disease which began to attack them in 1975.

It is greatly to our advantage that vegetative reproduction is so widespread in plants. Upon it depends much of our propagation of cultivated plants, especially of herbaceous and woody perennials, although annual and biennial crops of farm and garden are usually propagated sexually by means of seeds. One advantage derived by man from this vegetative reproduction is that mature plants can be obtained more quickly than by sexual means. Another is the uniformity of genetic constitution which it ensures, although, as in the case of the elm, this can result in high susceptibility to disease. The main devices used in vegetative propagation are the insertion of stem cuttings into suitable damp media, resulting in the development of adventitious roots; layering (exemplified in the natural tip layering already mentioned); and the grafting of a cutting (scion) onto a stock, the latter providing the root system and the base of the main stem for the resulting composite plant. All of these methods yield plants of predictable characteristics.

Sexual Reproduction in Algae

The argument that sexual reproduction must have been of early origin is supported by its presence, in simple form, in the unicellular algae, where it is well exemplified in *Chlamydomonas*, the adult vegetative form of which is haploid. This organism reproduces asexually by the division of the cell into two or four motile zoospores. Sexual reproduction may be effected by the division of the cell into eight motile gametes, but in some species, and in certain conditions, the vegetative adult may transform directly into a gamete without prior division. Such gametes are termed hologametes, those formed by division being termed merogametes. In either case, the gametes fuse in pairs to form diploid zygotes, which pass into a temporary resting phase, with thick cell walls. Ultimately, each zygote liberates four haploid motile cells, called meiospores, because they are formed after meiotic division. Each becomes an adult haploid *Chlamydomonas*. Such simple sexual processes have been elaborated in the evolution of higher organisms, in steps closely related to their exploitation of terrestrial life. We shall describe here only a few of these, sufficient to exemplify the principles involved.

One of these principles is the differentiation of the gametes into two types, distinguishable physiologically because members of one type will only fuse with members of the other type. Often, but not always, the two types are distinguishable morphologically as well, in which case they are termed anisogamous; if not so distinguishable, they are said to be isogamous. The two types, referred to as + and −, are called male and female by analogy with the sexual differentiation of higher organisms. If the two types are produced by two corresponding parental types, these also are termed male and female; the species is then said to be dioecious. Alternatively, both types of gamete may be produced by the same individual, which is then said to be monoecious or hermaphrodite.

Examples of such gamete differentiation are found in colonial members of the Phytomonadida, the order to which *Chlamydomonas* belongs. In *Gonium* (Fig. 8.3a), which forms a flat colony of up to 16 cells, each colony is capable of both asexual and sexual reproduction. Isogamous hologametes or merogametes are formed,

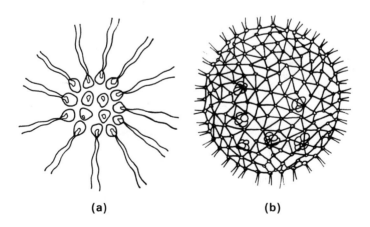

(a)   (b)

Fig. 8.3 (a) *Gonium*. (b) *Volvox* with six daughter colonies.

their behaviour showing them to be physiologically differentiated despite being morphologically identical. Each zygote yields four colonies, two of which are + and two −. In other genera, of which *Volvox* is an often-quoted example (Fig. 8.3b), there is morphological differentiation of the gametes, some individuals of the colony enlarging into macrogametes (large) while others divide to form microgametes (small). These two types, both of which are biflagellate, are often termed eggs (ova) and sperm respectively, again by analogy with higher forms.

Volvox also illustrates a tendency for the algal colonies to foreshadow the evolution of multicellular organisms, with some degree of cellular division of labour and of coordination. The colony, although built of many hundreds of separate individuals dispersed over the surface of a gelatinous sphere, swims by the coordinated movement of their flagella, with one pole of the colony foremost. The more anterior cells are smaller and sterile, while the posterior ones are larger and are capable of reproduction.

It has been suggested that true multicellular organisms might have evolved by the further differentiation and integration of such colonies, but this is speculation. There are, however, many other algae which, whatever their origin, have multicellular bodies (Fig. 3.1), and a more advanced sexual organization. One example is *Ulva* (Fig. 8.4), the common sea lettuce of the sea shore, which has an alternation of generations in which the two look identical (isomorphic). Its body is a green frond, anchored by a holdfast. The haploid generation produces biflagellate gametes at the margins of the blade by repeated mitotic division of cells, to be eventually liberated through small pores. Fusion takes place between gametes

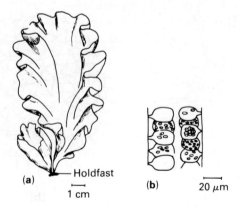

Fig. 8.4 *Ulva lactuca.* (a) Habit. (b) Transverse section of thallus producing zoospores. (From Bell and Woodcock, 1971.)

from different plants, showing them to be sexually differentiated (+
and −). The zygote divides to form a frond and holdfast which
look like those of the parent, but differ in being diploid and
reproducing asexually by producing zoospores (meiospores) from
marginal cells. Each meiospore eventually settles, divides, and grows
into a haploid plant which will reproduce sexually.

Other algae (the freshwater *Oedogonium* is an example) have life
histories similar in principle, but with macrogametes and
microgametes; the former may, as in *Oedogonium*, remain within its
parent cell; only the microgametes are motile, and they have to seek
out the macrogametes. The analogy with eggs and sperm thus
becomes even more marked than in *Volvox*.

It will be clear that what is happening in the stages briefly
outlined is that the diploid zygote, in contrast to that of
Chlamydomonas, is growing into an independent and fully developed
organism before it divides to form meiospores. It has been
suggested that this could have occured through a mutation
preventing zygotic meiosis, but whatever the explanation, the result
is the alternation of haploid and diploid phases which constitutes
the alternation of generations that we have already noted as characteristic of plants (p. 157). In *Ulva* these two phases are approximately equivalent in their contribution of adult organisms to the life
history; so they are in many other lower plants as well, but, as we
shall see, very different conditions evolved in the higher ones.

Terrestrial Life: Bryophytes

The earliest stages of the movement of plants onto land are
exemplified today by the Division Bryophyta, comprising the mosses
(Musci) and liverworts (Hepaticae), which are the simplest of the
terrestrial plants. As already seen (p. 47), they are characteristically
found in damp habitats. This is partly because they have little
capacity for resisting desiccation, but partly also because their
sexual reproduction requires a film of moisture sufficient to enable
the motile male gametes (antherozoids) to swim to the eggs.
Information regarding their origin is lacking, for no intermediate
forms are known, either surviving or as fossils, but they are
assumed to have originated from the Algae. Suggestive evidence for
this is the germination of their spores into a filamentous and alga-
like protonema which develops into the mature gametophyte.

Reproduction involves alternation of generations, which is
markedly heteromorphic; that is, the two phases are unlike. The
gametophyte in this instance is the conspicuous one, the sporophyte
being wholly or partly parasitic upon it. In the liverwort *Marchantia*
(Fig. 8.5) the gametophyte is a flattened thallus, dichotomously

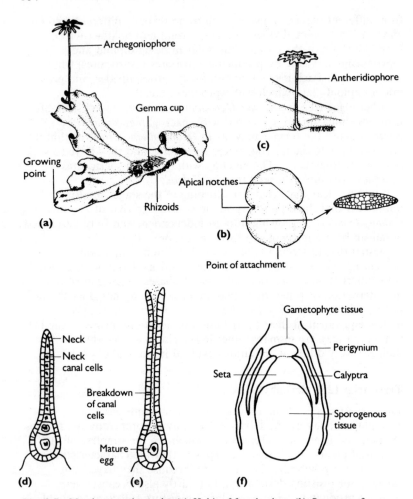

Fig. 8.5 *Marchantia polymorpha.* (**a**) Habit of female plant. (**b**) Structure of gemma.
(**c**) Antheridiophore. (**d**) Archegonium before breakdown of neck cells.
(**e**) Archegonium ready for fertilization. (**f**) Longitudinal section of sporophyte
rupturing the calyptra.

branched, and dioecious, the gametes developing on umbrella-like
structures (archegoniophores and antheridiophores) which arise at
the growing points of the thallus. The female gamete, a relatively
large egg, lies in a flask-shaped archegonium (characteristic of the
lower terrestrial plants and of many gymnosperms) at the base of a
long canal, down which the minute male gametes (antherozoids)
must swim to effect fertilization after breakdown of the canal cells.

The antheridiophores bear antheridia, protected within pits, each antheridium differentiating to contain large numbers of cells (antherocytes), in each of which there forms a biflagellate antherozoid. This consists of little more than an elongated nucleus, with some investing cytoplasm that probably contains mitochondria, so that the gametes are markedly anisogametic. This pattern of differentiation, in which the male gamete is primarily differentiated for transmission of a genome, is common throughout the animal kingdom, and also in those plants which retain such motile gametes. The diploid zygote of *Marchantia* differentiates into a sporophyte which remains attached to the gametophyte and is dependent on it for nutrition; it is attached by a foot and protected by gametophyte tissue (calyptra and perigynium). Spores are eventually formed, after meiosis, from sporogenous tissue, within a terminal spore capsule, and are distributed through the air to germinate into protonemata when they alight on damp surfaces.

Asexual reproduction takes place in *Marchantia* by fragmentation of the thallus or, alternatively, by dispersal through the air of highly specialized gemmae. These are multicellular bodies with two apical notches, each containing a meristem which gives rise to a new thallus; the two partners separate after degeneration of the central portion of the gemma.

Mosses have life histories similar in principle to that of *Marchantia*. The haploid gametophytes are either monoecious or dioecious, archegonia forming at the apices of the axes, while antheridia develop as club-shaped outgrowths on the same or other axes. Fertilization is brought about by the antherozoids swimming to the eggs, the zygote then differentiating into the diploid sporophyte, which carries a complex spore capsule. As in the liverworts, the diploid sporophyte depends upon the gametophyte for its nutrition, although it initially contains some chlorophyll and is in part self-supporting for a time. Meiosis leads to the production of haploid spores which are distributed in the air, and germinate on moist soil to form the gametophyte phase.

These bryophyte life histories have been briefly summarized because they point to the ways in which the problems of terrestrial life have been surmounted by the higher plants. The development of an inconspicuous sporophyte which is essentially parasitic upon the gametophyte is a particularly significant development, although, as we shall see, it has not been a universal one. The production of large numbers of spores for aerial distribution is another terrestrial feature, and one which permits rapid spread of colonization under favourable conditions. Nevertheless, bryophytes have no vascular tissue, nor true absorbing roots, the nearest approach to these being the rhizoids born by the thallus. The lack of these structures in bryophytes, in conjunction with their proneness to desiccation and their dependence upon water for effecting fertilization, limit the group to damp habitats. There is to some extent an analogy here with amphibians, which, as we have noted, appear in fossil deposits in the Upper Devonian, at about the same time as, or a little later than, the bryophytes. However, animal mobility, together with more complex morphological differentiation, has enabled amphibians, in certain special cases, to achieve what we have already seen to be a considerable measure of ecological diversification.

Lower Tracheophytes

The next major step in the terrestrial adaptations of plants was the emergence of tracheophytes, with differentiated conducting tissue. Although we know so little of the mode of origin and history of the early land plants, certain conclusions relevant to the evolution of tracheophytes can be drawn from biochemical considerations. First, the emergence of plants upon land would have exposed them to a much greater intensity of solar radiation, for, as we have seen, much of this is lost to aquatic organisms by reflection, while the remainder is absorbed in the upper layers of the water. The increased photosynthesis thereby made possible would have favoured differentiation and diversification of the plant body. On the other hand, algae release much of the products of photosynthesis into the water (up to 35 %, according to one estimate). This could not have occurred in fully terrestrial plants, which were thus in danger of accumulating large and possibly even toxic amounts of carbo-hydrate. It has been suggested that this would have imposed a need for metabolic adaptation, expressed perhaps in the appearance of thicker cell walls, and especially in their lignification, which are such important features of the structure of higher plants. The thought is an attractive one, for cellulose and lignin, being osmotically inactive, would remove the carbohydrate from interference with cell metabolism. However this may be, it is certain that the acquisition of conducting tissue, including the lignified xylem, was of crucial importance in permitting the efficient absorption and conduction of water and nutrients, together with developments in scale and complexity of organization which culminate in giant forest trees. Accompanying these developments were changes in the life cycle, permitting improved adaptation of the Tracheophyta to terrestrial conditions. Especially important has been the development of seeds, enclosing the embryo, together with food material, in a protective coat. This habit, which greatly reduces the vulnerability of the embryo, and makes possible provision for its dispersal, is characteristic today of the conifers and flowering plants;

its immense advantage is indicated by its establishment at a relatively early stage of plant evolution, in gymnosperms of the Carboniferous.

The lower tracheophytes, including the lycopods (e.g. *Lycopodium*, the club moss), the Sphenopsida (e.g. *Equisetum*, the horsetail), and Filicinae (ferns), are still archegoniate (that is, the ovum is still retained within the flask-shaped archegonium), and they still release antherozoids which swim through water to enter the neck of the archegonium. In these respects they resemble the bryophytes, but there is a fundamental difference in that the sporophyte is the conspicuous phase in the alternation of generations.

The plant body of a fern, for example, with its roots, fronds and vascular axis, is the diploid sporophyte (Fig. 8.6). The haploid spores, usually identical in appearance (homospory), are produced in stalked sporangia on the under surfaces of the fronds. After aerial distribution, they germinate on moist soil, each giving rise to a heart-shaped and haploid green (autotrophic) gametophyte (prothallus), which is anchored to the soil by rhizoids, and which bears antheridia and archegonia on its ventral surface. The zygotes develop into young sporophytes, the gametophytes disappearing when the new young plant has become sufficiently well established. In due course, the diploid sporophyte, which is perennial, releases meiospores and the cycle is completed. This is in large measure a truly terrestrial cycle. The ephemeral gametophyte, however, with its motile antherozooids, is still dependent upon water; an ecological limitation which has been overcome in gymnosperms and angiosperms by further specialization of the life cycle.

Gymnosperms

The principle of the life cycle is the same in both of these groups, but there is an important difference between them in that the seeds are exposed (*gymnos*, uncovered) in the gymnosperms, but protected within an ovary in the angiosperms (*angeion*, vessel). The gymnosperms form a comparatively small group of some 700 species, but within this group the conifers (e.g. *Pinus*) are of outstanding ecological importance, for their trees, which are the sporophyte phase, are widespread, forming the climax vegetation in cooler regions and on mountains. Even so, their intensive exploitation by man, combined with the effects of fire, disease and insect pests, is demanding enlightened and worldwide reforestation to ensure, in conjunction with improved management techniques, their survival in amounts adequate for our future demands.

The meiospores of the pine are of two sorts (heterospory), both

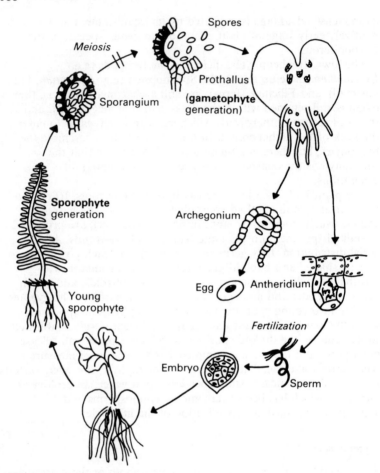

Fig. 8.6 The life cycle of a fern.

being formed on specialized structures, the cones (Fig. 8.7). Smaller microspores developed in microsporangia born on modified leaves (microsporophylls) which form the male cones. These develop in clusters at the tips of the previous season's growth. The megaspores develop in megasporangia on modified dwarf shoots which constitute the female cones; these arise on laterals of the current season's growth. The microspore nuclei divide to form four-celled microgametophytes, and it is these, winged for aerial distribution, which are distributed as yellow pollen grains in the spring. The female cone bears ovules, each of which consists of a protective

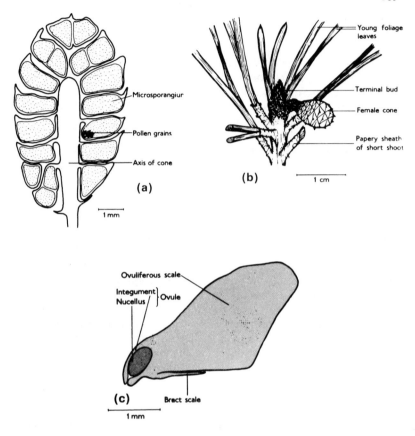

Fig. 8.7 *Pinus sylvestris*. (a) Longitudinal section of male cone. Each microsporangiophore bears two pollen sacs. (b) Female cone in the early summer of its first year. (c) Longitudinal section of a single scale. (From Bell and Woodcock, 1971.)

integument enclosing specialized tissue, the nucellus, within which a megaspore parent cell differentiates. Subsequent events are complex, and long drawn out, and only the salient ones can be mentioned here. Following the arrival of the pollen (pollination), the megaspore parent cell divides by meiosis, and this initiates the formation of the female haploid megagametophyte. In due course this develops one or more archegonia which are eventually reached by pollen tubes formed from the pollen grains. One of the original four cells of the microgametophyte (tube cell) is probably responsible for promoting the tube development. Another of the four (generative cell) divides to form a stalk cell and a body cell, the latter dividing again to form two microgametes (sperm). One of the sperm nuclei fuses with the egg nucleus, thus forming the zygote which initiates the diploid sporophyte phase.

The zygote divides to form an embryo, which takes its nutriment from the endosperm, this being gametophyte tissue in which food material has been deposited. The integument forms a very hard seed coat protecting the embryo (sporophyte), the whole forming the seed. By this stage the female cone has become woody, its scales open out, and the seeds, which are winged for dispersal, are distributed through the air in the autumn of the year following that in which pollination occurred. Germination is initiated by the embryo breaking through the seed coat.

The advances made by the gymnosperms are not difficult to appreciate. Heterospory, already found in some highly specialized ferns, is general, and has led to the production of reduced micro- and macrogametophytes which are no longer free-living. They are completely dependent upon the sporophyte, in contrast to the gametophytes of ferns, which are free, usually autotrophic, and dependent upon moist habitats. Further, water is no longer needed for fertilization, and the zygote develops into an embryo which is provided with ample food reserves protected within a seed coat. Yet these adaptations, adequate though they have proved for terrestrial life, were capable of further development, for out of them evolved the more advanced life cycle of the angiosperms (flowering plants).

Angiosperms

These, with some 200 000 species, immensely surpass in diversity the relatively few species of gymnosperms. Moreover, they probably did not appear until the end of the Jurassic or the beginning of the Cretaceous, yet within some 25 million years (during the latter period) they rapidly evolved to dominate terrestrial vegetation, and to enter into a closely integrated relationship with certain groups of animals. This spectacular success calls for explanation.

Sexual reproduction in flowering plants follows the principles established in the gymnosperms, but with important advances and refinements, full details of which will be found in botanical texts. One crucial development is the flower (Fig. 8.8), which is a specialized stem that can, in principle, be derived from the cone of the pine, except that both microspores and megaspores develop on sporophylls, and are usually carried on the same flower. This organ is typically composed of four groups of leaf-like structures; sepals, forming the calyx, at the base; next the petals, forming the corolla; then the stamens (microsporophylls), consisting of filaments bearing

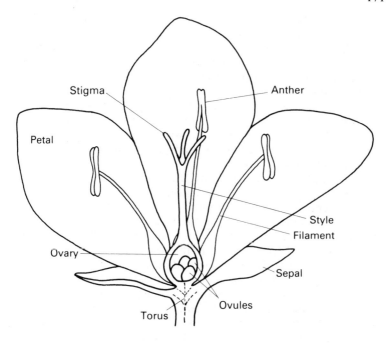

Fig. 8.8 Diagram of a generalized dicotyledonous flower in longitudinal section.

the pollen grains in terminal anthers; and finally the pistil, composed of one or more leaf-like carpels (megasporophylls). The pistil comprises a basal ovary, with one or more ovules, and a slender style which expands at its tip to form the stigma.

Heterospory leads, as in gymnosperms, to the production of reduced megagametophytes and microgametophytes, but these are of simpler structure. The haploid microgametophyte (pollen grain), following fertilization, forms a pollen tube, consisting of a single cell, down which pass two sperm (microgametes). The haploid megaspore (the surviving one of a group of four), enclosed within the wall of the megaspore, develops into the megagametophyte (embryo sac). This consists of six cells, one of which is the egg, lying near the micropyle with two synergid cells. The other three cells (antipodals) are at the opposite pole, while two free nuclei (polar nuclei) lie centrally. The nucellus containing the embryo sac is surrounded by integuments with an opening, the micropyle, the whole constituting the ovule (megasporangium).

The pollen tube penetrates the micropyle, and the two sperm enter the embryo sac. Subsequent events are very characteristic of flowering plants. One of the sperm nuclei fuses with the two polar nuclei to form a triploid nucleus, the primary endosperm nucleus, while the other sperm nucleus fuses with the egg. The zygote divides to form a developing diploid embryo, while division of the primary endosperm nucleus leads to a cellular structure in which food reserves are laid down to form the endosperm. The remaining five cells of the embryo sac disappear. The resulting complex is the seed, protected within a seed coat developed by modification of the integuments. Metabolically inactive, with a very low water content, it can remain dormant for long periods, during which it is highly resistant to adverse conditions.

These, in essence, are the events of angiosperm sexual reproduction (Fig. 8.9), but they depend for their successful operation upon further adaptations. The immediate stimulus to fertilization, following pollination, is the action upon the pollen grain of a secretion of the stigma, but this requires that the pollen should first have been conveyed to the stigma to effect the pollination. A few angiosperms depend on water for this, but many, like gymnosperms, depend upon the wind; oaks and grasses are examples of the latter. Many make use of insects (including butterflies, moths, flies and, especially, bees). They are attracted to the flowers by the conspicuous colours of the corolla, or of the calyx (or, exceptionally, of modified leaves or bracts, as in *Poinsettia*), and by the presence of nectar, a sugary fluid secreted by glands called nectaries. Thus they pick up pollen (itself a food for bees) which they then transfer to other flowers. Floral structure, together with other adaptations such as the production of distinctive scents, ensure that the visiting insects are not only attracted to the flowers, but also become dusted with pollen. All of this makes it evident that the evolution of angiosperms has been closely integrated with the evolution of insect structure and behaviour, and not least with their powers of photoreception and chemoreception.

Further integration results from another characteristic of angiosperms: the development from the ovary of a seed-bearing structure, the fruit, typically with a pericarp formed from the wall, but sometimes with the involvement of other parts as well. Fruits increase immensely the successful distribution of the seeds, sometimes by wind action, but often by animals. Some fruits are dry when fully ripe, an example, associated with aerial distribution, being the monocarpellate legume of peas and beans, which splits (dehisces) as it dries, so that the seeds are thrown from the parent plant. Others are the capsule of poppies, formed from multicarpellate pistils which split open so that the seeds are distributed by shaking, and the indehiscent samara (found in the ash,

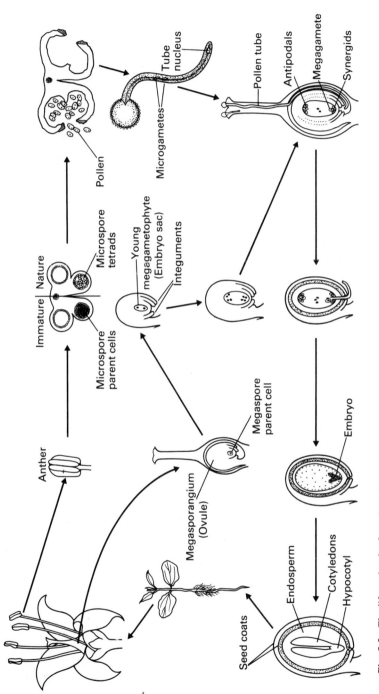

Fig. 8.9 The life cycle of a flowering plant.

for example) which possesses wing-like projections that ensure wide aerial distribution.

Animals become actively involved when the ripe fruit has a soft and attractively flavoured texture. Common examples are berries, consisting of one or more seeds enclosed by a fleshy pericarp. Drupes have the inner layer of the pericarp hard, forming a 'stone' surrounding the seed (e.g. cherry), while in 'false fruits' parts of the flower additional to the carpels contribute. In the apple, for example, the 'core' of the fruit is the inner layer of the ovule wall. In all of these cases, the seed coats resist action of the digestive juices of the animals that eat the fruit, and are thus widely distributed. To be eaten, however, is not an essential condition for seed distribution by animals. Acorns and nuts may be removed and then dropped by squirrels and jays, for example, while burrs and other hooked fruits may be carried on fur.

All these adaptations ensuring the production, distribution and survival of the embryo have undoubtedly been major factors in securing the success and dominant ecological position achieved by flowering plants, but there is still another to be mentioned. This is that, the life history of angiosperms, given good conditions, can be completed very quickly, in marked contrast to the long-drawn-out reproductive processes of the gymnosperms. We have seen an extreme illustration of the ecological value of this in ensuring the survival of flowering plants in desert conditions, where they are able to profit from very brief periods of rainfall. More generally, however, the advantage of this speed is that it ensures, in less extreme conditions, a rapid sequence of generations. This means that gene mutations and recombinations will spread much more quickly in angiosperms than in gymnosperms, and this goes far to account for the speed of angiosperm evolution.

Full exploitation of the genetic advantages of angiosperm life histories depends, however, upon another set of adaptations. Cross-fertilization is needed to achieve the full advantages of genetic recombination, and this demands cross-pollination; that is, the carrying of the pollen from one plant to another. Actually, many angiosperm species are self-fertilized, sometimes because the pollen ripens when the stigma is fully receptive, or before the flower is fully open, but cross-pollination is probably the more common method.

There are various ways in which this is contrived. Sometimes the pollen ripens before the stigma of its own flower is receptive, although this does not eliminate the possibility of it fertilizing other flowers on the same plant. More effective is self-incompatibility, in which fertilization only takes place when the pollen alights on a different plant of the same species. Various devices are responsible

for this. It may, for example, be brought about by the stigma failing
to supply the appropriate stimulus for pollen tube growth, or there
may be some interaction between the tube and the style. When
insect transmission is involved, the structure of the flower may be
elegantly adapted to ensure cross-fertilization. In extreme cases, two
different types of flower may be produced, with only one type on
any one plant, and in these the positions of the stamens and stigma
may be reversed, so that pollen taken up on one visit must
necessarily be transferred to the stigma during a visit to a flower of
the opposite type. An example is the differentiation of thrum-eyed
(short styles and high anthers) and pin-eyed (long styles and low
anthers) flowers in *Primula vulgaris*. We have referred to this earlier
as illustrating the action of a supergene (p. 19).

9 Life Histories (continued)

Protozoans

Surprisingly, in view of the evidence for the early origin of sexual reproduction, this process seems to have been suppressed in many protozoans. Strictly, this could mean simply that investigators have failed to detect it in these forms, but in some widely studied species it certainly does seem to be absent. A familiar example is *Amoeba proteus*, although there is in the literature one report, never confirmed, of a possible sexual process. Another example is the flagellate protozoan responsible for human sleeping sickness, *Trypanosoma*, which is believed to be haploid. It has been thought that it may be primitively asexual, instead of having evolved from sexually reproducing ancestors by the loss of meiosis, but this is pure speculation.

These, of course, are special cases, for sexual reproduction is virtually universal amongst animals, as it is in plants. Its evolution, however, and its influence upon animal life histories, have followed along lines very different from those characteristic of the plant kingdom, in part because the mobility of animals has provided them with much greater scope for adapting to environmental problems, whether in water or on land. A fundamental difference, already emphasized (p. 156), is that the haploid state in animals is restricted to their gametes, the diploid zygote developing by diploid mitoses, with certain exceptions, some of which will be mentioned later. Alternation of generations does occur in animals, but its nature serves only to emphasize the difference between animals and plants in this respect. Two illustrations will make this point.

One concerns marine protozoans of the order Foraminiferida, amoeboid organisms that inhabit single- or multi-chambered shells which are often calcareous, and which are deposited as Globigerina ooze (named for a characteristic genus) on the Atlantic floor. They provide fossils that are abundant in chalk rocks and that are much used by the oil industry in the identification of strata. There is variation in the life histories, even of this one group, but some do show an alternation of generations that is similar in principle to that of gymnosperms and angiosperms. *Elphidium crispum* (= *Polystomella crispa*) is an example (Fig. 9.1), with a multi-chambered and spirally coiled shell. Flagellated isogametes are

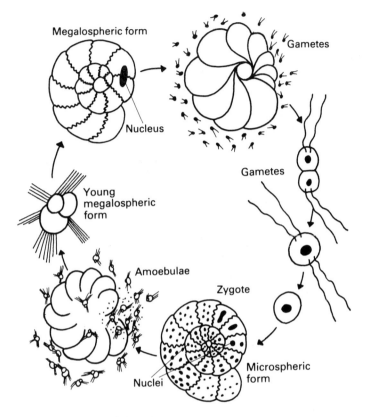

Fig. 9.1 Life cycle of *Elphidium* (*Polystomella*).

produced by haploid individuals (gamonts, in zoological
nomenclature). Fertilization, which takes place in the sea, yields
zygotes which grow into diploid individuals (agamonts). These
reproduce asexually by repeated division, with meiosis; the
amoebulae thus formed grow into the haploid gamonts. The cycle
with its alternation of asexual and sexual phases, and with meiosis
taking place at an intermediate stage, is unique amongst animals,
and emphasizes the difference between these and plants. It involves
dimorphism of the shell, the initial chamber (proloculus) being
smaller in the agamont (microspheric form) than in the gamont
(megalospheric form). Strangely, a related genus (*Spirullina*) has the
reverse dimorphism, with the smaller proloculus in the gamont.

This type of alternation of generations is not found in other
protozoan groups, but there is still much diversification in their life

cycles, evidenced particularly by some of these animals being haploid in the adult while others are diploid. We have mentioned *Trypanosoma* as an exclusively haploid asexual form. The endoparasitic telosporean sporozoans are also haploid, with an alternation of sexual and asexual phases, in which the zygote divides by meiosis immediately after its formation (zygotic meiosis). Examples are *Monocystis*, common in the seminal vesicles of the earthworm, and *Eimeria stiedae* (Fig. 9.2), a parasite of the hepatic cells of the rabbit. In *Monocystis*, multiplication occurs at only one stage, when each zygote divides to form eight sporozoites (sporogony). A similar phase occurs in *Eimeria*, but in addition there is multiplication by schizogony, the adults dividing to form numerous individuals (merozoites); these infect fresh cells where they may repeat the schizogony before producing gametes.

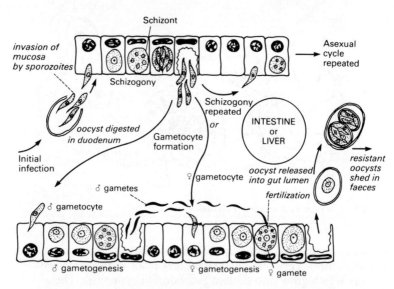

Fig. 9.2   Life cycle of a coccidian (e.g. *Eimeria stiedae* from the rabbit). (From Smyth, 1962.)

Other protozoans are diploid, examples being various flagellates and the Ciliophora (e.g. *Paramecium*). The life cycles of ciliophorans are complicated by the differentiation of the nuclear material into a highly polyploid macronucleus and a diploid micronucleus. Reproduction in these animals is asexual, by binary fission, but a reduced sexual process occurs in which two organisms (conjugants) temporarily associate and exchange haploid division products of the micronucleus, with subsequent fusion of these to form, in each individual, a new diploid nucleus, which then divides. The macronucleus degenerates and is replaced from the new micronuclear material, the two individuals having meanwhile separated, as exconjugants, and then divided.

Remarkable variation is found in the complex polymastigote and hypermastigote flagellates, many of which live in the gut of wood-eating termites (Fig. 4.15). *Macrospironympha* is a diploid form, with meiosis occurring when gametes are formed, while *Trichonympha* is haploid, with zygotic meiosis.

It is not easy to account for these and other variations in protozoan life cycles, although some experimentation and variability is to be expected in organisms that stand taxonomically so close to the first animals. It would seem reasonable to suppose that each group or, perhaps, species would have evolved a pattern of reproduction appropriate to its mode of life, some balance being struck between the speed and genetic stability of asexual reproduction, and the variability associated with the sexual method. It has been argued that haploid organisms could be capable of rapid adaptation because mutant genes will not be masked by normal dominant alleles, while diploid forms must benefit by the testing of the mutant genes in different genetic environments. However, there is little assured basis for such interpretation. The variability within the flagellates remains difficult to explain, unless it be, according to one suggestion, that the sheltered environment of an insect alimentary tract, with ample food supply, encourages diversification, or at least permits variants to survive that might not be able to do so in more rigorous environments.

Alternation of Generations in Metazoans

The life cycles of metazoans, organized around the principle of diploid organisms with haploid gametes, have provided ample scope for ecological adaptation. An alternation of sexual and asexual phases, already noted in protozoans, is particularly characteristic of coelenterates, where it is often associated with marked polymorphism. *Obelia* (Fig. 9.3), for example, has a dioecious free-swimming stage, the medusa (jelly fish), which alternates with a sessile polyp; this gives rise to a polymorphic colony by asexual reproduction, the daughter individuals remaining attached. The medusa, which is adapted for planktonic life, and which is produced as asexual buds of the colony, produces haploid gametes by meiotic division. The ova are fertilized in the sea by the sperm, which are of the flagellate type common in animals, and develop into a ciliated larva, the planula. This settles and grows into a new colony,

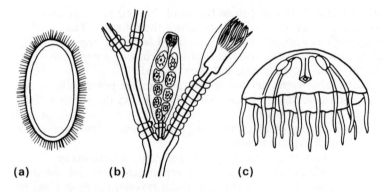

Fig. 9.3 (a) Planula larva. (b) *Obelia*, part of colony showing a hydranth and medusae developing on a gonophore. (c) *Obelia* medusa.

composed of two types of polyp, feeding hydranths and reproductive blastostyles, the latter budding off the medusae to complete the cycle. Polymorphism is here shown within the colony, and also between the polyp and medusa.

The function of the motile medusa in the life history is sufficiently obvious; it provides for distribution and colonization of new habitats. Yet to say this is to leave some unanswered questions. One arises from the widespread tendency in the coelenterates for the reduction and elimination of the medusa. The solitary polyp of hydra is an extreme example of this, its budding producing only polyps like itself. Here the loss of the medusa (and there is taxonomic evidence that hydra must have evolved from a stock with alternation of generations) can be regarded as an adaptation to freshwater conditions; these often lead to the loss of motile developmental stages because of the danger of them being swept away into unsuitable habitats, or even out to sea. But most coelenterates are marine (hydra being a secondarily freshwater form), and for them the explanation of the reduction of the medusa stage is not so clear. One must suppose that here, as so often, the pattern of life cycles is determined by a balance of advantages. Perhaps natural selection has favoured reduction of the medusa because this stage proved vulnerable to predation in the plankton, and because distribution of the colony can be adequately secured by a brief excursion of the planula larva.

An alternation of sexual and asexual generations is rare elsewhere in the animal kingdom, and its supposed occurrence sometimes rests on a misinterpretation. One example of this is the life history of the liver fluke, *Fasciola*, the trematode (flatworm) parasite of the bile ducts of sheep (called the primary or definitive host because it is

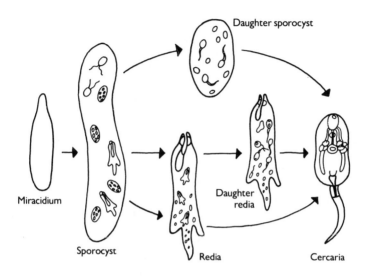

Fig. 9.4 Pattern of larval development in trematodes. (From Smyth, 1962.)

here that the parasite reaches sexual maturity). This life history also illustrates the way in which many parasites produce large numbers of developmental stages called larvae (p. 187) to ensure transmission to another host, often, as here, by transmission through a secondary or intermediate host.

The fertilized egg of *Fasciola* yields a miracidium larva (Fig. 9.4) which enters a snail, and develops there into a sac-like sporocyst filled with propagative cells and developing larvae of a new type (redia larvae). The sporocyst may produce a second generation of sporocysts, while the redia larvae, liberated from these, produce more redia larvae before developing into yet another larval type, the cercaria. This leaves the snail and can then infect a sheep when this eats grass on which the cercaria has encysted. Virtual certainty of successful infection of the vertebrate host is ensured by the exploitation, as secondary host, of an amphibious snail which lives in damp meadows but which can resist desiccation, and also by the extraordinary multiplicative power shown by the parasite in its life cycle. The lesson is to control the parasite by adequate drainage of the meadows in which the sheep are feeding, for this removes the snail. It is estimated that in the absence of such measures, a single egg could give rise to as many as 1 000 000 cercariae, an achievement in multiplication that is unsurpassed by any other metazoan group.

This process, however, is not one of asexual reproduction, nor does it involve alternation of generations. The preferred interpretation is that all of the larvae develop from germinal cells derived direct from the initial ovum. This life cycle is therefore to be regarded as an exploitation of polyembryony, which is the formation of several embryos from one ovum. It is not confined to these parasites (human identical twins are another example of it), but trematodes certainly use it in a highly specialized way, well adapted to ensuring transmission.

In contrast to the above, certain other flatworms (cestodes, formed of chains of hermaphrodite reproductive units called proglottids) do have a phase of truly asexual multiplication in their life cycle (Fig. 9.5). One example is the tapeworm *Multiceps multiceps*, parasitic in the alimentary tract of certain carnivores, including the dog and the fox. In its intermediate hosts, which are ungulates, and particularly sheep (a few cases have also been reported in man), it gives rise to bladder worm stages in the brain and spinal cord, producing a disease called 'staggers'. The life cycle is completed when the infected organs are eaten by the primary host. It is usual in such tapeworm cycles for the bladder worm stage, or its equivalent, to give rise to only one tape worm, but in this species the bladder (here called a coenurus) buds off many hundreds of tape worm heads (scolices) from its inner wall, so that a very severe infection can develop in the primary host.

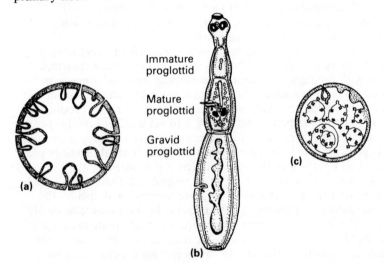

Immature
proglottid

Mature
proglottid

Gravid
proglottid

(a)

(b)

(c)

Fig. 9.5 (a) Coenurus larva of *Multiceps multiceps* showing budding of scolices from inside wall. (b) *Echinococcus granulosus*. (c) Hydatid cyst of *E. granulosus* showing formation of brood capsules. Not to scale. (From Smyth, 1962.)

Even more spectacular is the life history of another tape worm, *Echinococcus granulosus*, a small organism (Fig. 9.5) parasitic in the intestine of carnivores, particularly of dogs. The bladder worm stage (hydatid cyst), found especially in sheep and cattle (although man is not excluded), grows to a huge size, budding off internally a number of brood capsules in which many scolices form. The result may be a 50 cm bladder containing up to 15 million tapeworm heads. Control of this efficiently organized parasite can, however, be secured in endemic areas by rigorous hygiene in the handling of dogs.

Asexual reproduction occurs in some freshwater oligochaete worms, presumably correlatal with the considerable powers of regeneration (replacement of lost parts) possessed by this group. The possibilities are illustrated by *Nais* (Fig. 9.6). *N. paraguayensis*, in addition to its sexual reproduction, breaks up into a number of pieces, each of which regenerates into a complete worm. This process, termed fragmentation, is reminiscent of the vegetative reproduction of angiosperms; it is said to result in some 15 000 worms arising in two months from one parent. This mode of reproduction provides for rapid increase in numbers with some economy of effort, since it does not require the production of eggs; indeed, the process may continue indefinitely, at least under laboratory conditions, without sexual forms appearing at all. Its value for freshwater animals is not clear, although it perhaps provides for rapid colonization of small or temporary bodies of water during briefly favourable conditions.

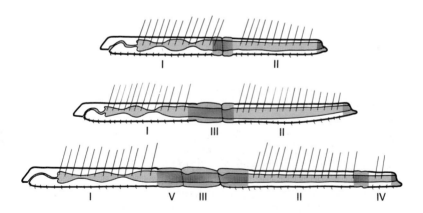

Fig. 9.6 Asexual reproduction of *Nais*. Chains of 2, 3 and 5 are being formed, the roman numbers showing the order of production. The shaded regions, in which chaete (bristles) have not yet appeared, represent next tissue. (From Stephenson (1930). *The Oligochaeta*. Clarendon Press, Oxford.)

Parthenogenesis

A variant of normal sexual processes, restricted in distribution but ecologically important when it does occur, is parthenogenesis, which is the development of ova without fertilization. Well-known examples are the rotifers. These are minute but very abundant animals, predominantly freshwater in habit (Fig. 9.7), feeding by means of a ciliated band, the 'wheel', and hence called 'wheel animalcules'. In some species the male is unknown, but in others the mode of reproduction is adapted to the season, and to conditions in ponds and streams, as we have already suggested for naid worms.

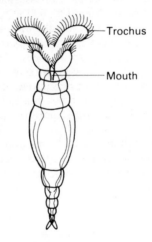

Fig. 9.7   A rotifer. These animals range in length from 100 to 500 μm.

Rotifer eggs that are to develop parthenogenetically are diploid; they cannot be fertilized, even if males are available, they are laid in thin shells, and they hatch in a few days. These parthenogenetic eggs, which are produced in the spring and early summer, provide for rapid multiplication. Later in the year, when the population becomes overcrowded through the drying up of the habitat, or perhaps in response to other environmental signals, haploid eggs are produced. Those which are not fertilized develop into males. Others are fertilized by these males, and form diploid zygotes, which contain yolk reserves, and are enclosed in thick shells. These are dormant stages, which can resist adverse winter conditions. Eventually they hatch into females which reproduce by parthenogenesis.

This type of life history, which permits variations adapted to particular types of habitat, provides for rapid and economical exploitation of a fluctuating freshwater environment. The same is true of many branchiopod crustaceans (e.g. *Daphnia*) which have parthenogenetic life histories of a similar type, culminating in the production, by fertilization, of dormant eggs enclosed in a thick protective capsule called the ephippium.

The outstanding illustration of the value of parthenogenesis in exploiting an environment with marked seasonal variation, and in relation to severe selection pressure, is found in aphids. These small, highly diversified and extremely abundant insects have attracted close interest primarily because of their incidence as pests of cultivated crops, but they are of importance also as foci of pressures exerted by a variety of parasites and predators, most of which are themselves insects. Estimates indicate that 2260 million individuals may be feeding at any one time on the roots, shoots and leaves of one acre of ground. This figure, taken in conjunction with a further startling estimate that one aphid could, in theory, give rise in one season to progeny equal to the weight of 10 000 men, speaks well for the efficiency of a reproductive technique which is closely integrated with the habits of angiosperms. These insects arose in the Carboniferous or early Permian, and it is significant that much of their diversification seems to have occurred much later, after the appearance of those plants. (We have noted earlier the elegant adaptation of their suctorial feeding mechanism to the location of the sieve tubes in the vascular bundles.)

A relatively simple type of aphid life cycle (Fig. 9.8) is exemplified by *Drepanosiphum platanoides*, which lives exclusively on the sycamore tree, *Acer pseudoplatanus*. Dormant eggs which have survived the winter hatch in the spring, yielding winged (alate) females which are viviparae, so called because they produce viviparously (by live birth) a series of parthenogenetic generations. These viviparae fly from one tree to another when stimulated to do so, by overcrowding. At the beginning of autumn there is a change of reproductive phase; the eggs now yield wingless (apterous) females which produce eggs (oviparae) and winged males which mate with them; from this mating arise the overwintering eggs.

More complicated life cycles than this are common. For example, aphids may alternate between different hosts (between trees and grasses, for example), some may never produce normal sexual forms, while others may produce two sorts of clones, one exclusively parthenogenetic and the other including normal sexual forms.

The changes in phase of the cycle are determined primarily by environmental signals, including temperature, photoperiod, the condition of the host and the quality of the food which it provides.

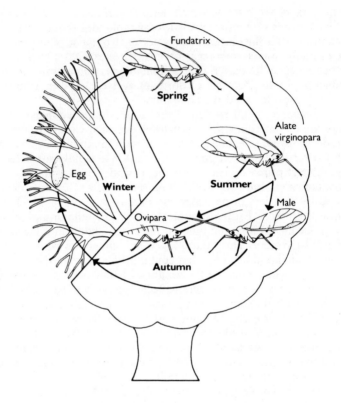

Fig. 9.8 Life cycle of the sycamore aphid. (From Dixon, 1973.)

For example, oviparae appear in the sycamore aphid in early
September, when the photoperiod has shortened, the females
appearing first and the males later. In other species, low
temperature has been shown to be a factor, while another is the
concentration of amino nitrogen in the host's sap, this being
relatively low in summer and high in early spring and in autumn
(p. 64). In addition to these exogenous factors, the responses of
aphids are also influenced by their own internal conditions, which
show cyclical changes. This is evident from the fact that in the early
part of the year the insects are much less influenced by photoperiod
than they are in autumn.

The complexity of the controlling factors in these aphid cycles is a
measure of the high selective advantage which must have
determined their evolution. Aspects of the advantage are not far to
seek. The apterate condition is an economy, for the resources which

would otherwise be used for growing wing muscles are made available for reproduction. Thus it has been noted that certain individuals with shorter wings, and hence with reduced wing musculature, are more fecund than those with fully developed ones. Other advantages are that the association of parthenogenesis and viviparity make for an economy in time which facilitates very rapid multiplication; the appearance of alates permits the avoidance of overcrowding and poor nutritive conditions in the host plant, while the production of overwintering eggs by the normal sexual phase contributes to survival during adverse conditions. No group of animals illustrates better the subtlety with which animal life cycles can be adapted to complex environmental demands.

Larvae

One of the most important of the advantages which animals gain from their motility is the possibility of extending and elaborating their life cycles by the insertion into them of a larval stage. We have touched on some examples of this already, but we return to it now in more detail because so many groups of animals have made use of this adaptation in exploiting their environment. A larva is a developmental stage which leads its own independent life, usually a free-moving one. It differs markedly in structure and habits from the adult, so that in order to develop into this it has to undergo a metamorphosis which may be very drastic. This mode of development is termed indirect, contrasting with direct development, in which the adult form is attained by a progressive growth and differentiation from the zygote.

Several advantages accrue from the possession of larval stages. Firstly, they provide for dispersal, comparable to that secured by the production of spores and seeds by plants. We have already seen how important such dispersal is for sessile forms and for parasites (p. 180), but it is also valuable for free-living animals that occupy restricted habitats. Bottom-dwelling worms and crustaceans, for example, may well be able to extend their range more effectively through their larvae than through their adults. Such dispersal not only overcomes risks of overcrowding and consequent reduction of food supplies; it also discourages close inbreeding, with the genetic and evolutionary advantages that we have seen to flow from this.

A second advantage results from young stages often requiring conditions of life, and especially types of food, different from those that are suitable for the adults. Indeed, it is often feeding habits more than anything else that underly the differences between larval and adult organization. It follows that larvae are very far from being simple and unspecialized organisms. On the contrary, their independence, and their individual exploitation of the environment, expose them to selection pressures that may be very different from those acting on the adults. Often, therefore, they become specialized quite differently from the adults, and it is this that makes drastic metamorphoses necessary. Insects provide extreme examples.

The development of these animals, like that of other arthropods, involves a series of moults (ecdyses), in which the outer hard exoskeleton is shed and a new one formed to accommodate the larger body. The stage between two successive moults is called an instar. This mode of development, which is a necessary consequence of the restraint on body form imposed by the exoskeleton, is expressed in two main patterns. In the more primitive insects, exemplified by cockroaches, the young resemble the adults in many features, differing chiefly in lacking fully developed wings and sexual organs. Their metamorphosis is, in consequence, relatively slight, and at each instar they tend to approach increasingly closely to the adult form. For these reasons, the young stages are sometimes called nymphs, although it has been argued that they should be considered as larvae. Insects with this type of life history are called Hemimetabola (Heterometabola, Exopterygota), the last of these terms referring to the progressive development of the wings as external structures.

More advanced insects have larvae which specifically lack the compound eyes that are present in nymphs and adults, but which differ from the adult in many other features as well. So much so that there is interposed between larva and adult a pupal stage, in which extensive reorganization of tissues takes place. Butterflies and moths are familiar examples. Thus the larva of the giant American silkworm moth (*Hyalophora* (*Platysamia*) *cecropia*), adapted to function as a formidable herbivore, increases its weight some 5000 times during early summer. It then pupates, and from the pupa emerges an adult moth which is adapted to function in flight and reproduction. Such insects are called Holometabola.

The possibilities of diversification within the general pattern of indirect development are multitudinous, and we can here touch upon only a few examples.

Certain marine polychaete worms have a ciliated, unsegmented and biconical larva (trochophore, Fig. 9.9), which lives for a time in the plankton, eventually settling and growing into the adult worm, with the progressive formation of segments bearing the characteristic parapodia and bristles (chaetae) which contribute to adult locomotion. But even here independent larval evolution has led to variants. Such a one is the nectochaetous larva, also common in the plankton, in which the newly hatched larva already possesses some chaetigerous segments.

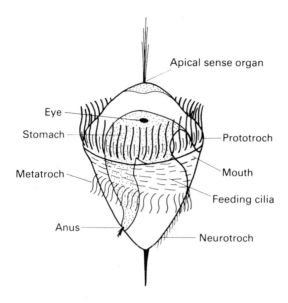

Fig. 9.9 Trochophore larva of *Pomatoceros triqueter* about 6 days old. Length of body about 200 μm. The prototroch and metatroch are characteristic ciliated bands used for locomotion.

Altogether more complicated are the life histories of crustaceans (Fig. 9.10). The simplest of their larvae is the nauplius, an organism with three segments, each bearing a pair of limbs. These, however, differ from the form of the adult limbs, being adapted for larval locomotion. Growth, accompanied by a series of moults, may produce a series of increasingly large nauplius stages, as happens in copepods. The nauplius stages of these animals, however, are eventually succeeded by a series of copepodid stages, which approximate increasingly closely to the adult. In such cases metamorphosis is a long-drawn out process, but special circumstances can result in a much more drastic one. This is so in certain highly modified parasitic copepods, where the larvae, by remaining free-living, have necessarily diverged considerably from their adults. Thus *Salmincola salmonea*, parasitic on the gills of the Atlantic salmon, *Salmo salar*, has a copepodid larva which attaches to the gills of the fish when this enters fresh water on its spawning migration. A series of moults follows, during which copulation takes place, followed by death of the male parasite and permanent attachment of the female, which develops large egg sacs.

Drastic metamorphosis may also be imposed by sessile life, notably in barnacles (Cirripedia). The crustacean nature of these

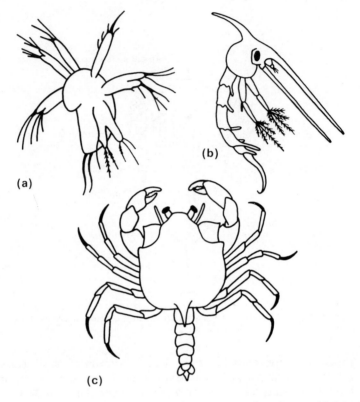

Fig. 9.10 Some representative crustacean larvae. (**a**) Nauplius of a penaeid shrimp. (**b**) Zoea of a crab. (**c**) Megalopa of a crab.

gregarious animals, which is by no means obvious when first inspected on the sea shore, is sufficiently apparent when their crustacean-type limbs emerge for filter feeding from the valved shell. The life history of cirripedes begins with a nauplius of distinctive form. Later this becomes a bivalved cypris larva which settles and metamorphoses.

More advanced crustaceans, and especially the Decapoda, have life cycles which are both more complex in their larval adaptations for independent life, and yet also shortened, in that the nauplius stage is passed through in the egg and the larva hatches in a more advanced form. Thus crabs (Brachyura) hatch as segmented zoea larvae, already with a well developed and segmented thorax and abdomen, and with some thoracic limbs; closer to the adult crustacean form than is the nauplius. Later, the zoea stage is succeeded by a megalopa larva, a crab-like form with all of its appendages present. It still swims for a time in the plankton, but eventually sinks to the sea bottom and completes its metamorphosis.

The apparent simplicity of these invertebrate larvae is deceptive. If the species is to survive, some larvae must succeed in settling at points where the adults can live, and this is no blind process. Detailed study of certain larvae (that of the polychaete *Ophelia*, for example, or of the barnacle) shows that the larvae of benthic adults explore the substratum before finally settling and metamorphosing, remarkable sensory discrimination being involved. During the search they are sensitive to many stimuli. These include currents, illumination, the nature of the substratum, the size of its particles, the microscopic flora and fauna living on and between those particles, and, for barnacle larvae, the ability to recognize chemically their own species, and to distinguish between it and related ones. Chemical communication (p. 222) is certainly an important factor, linking the larvae with biological characteristics of the type of substratum that will suit the adult. Here is an example of the close interweaving of members of a community, dependent in this instance upon remarkable specializations of the larvae which go far beyond their more obvious locomotor adaptations.

Viviparity

We have already noted in aphid life cycles the alternation of viviparity and oviparity, but viviparity occurs in other groups as well, and we must now consider it in a wider context. In general, it helps the young during their early development, and to this extent there is a slight analogy with the retention in plants of the female gametophyte within the parental sporophyte, for this also has a protective function. We have seen that this retention has contributed to the exploitation by plants of terrestrial life. So has viviparity in animals, but the comparison cannot be pressed very far, for there is nothing in the life processes of seed plants to match the provision in animals of maternal nutriment sufficient to produce a fully developed and functioning offspring prior to birth.

Nutritive dependence is one of the characteristics of viviparity, and is the basis for a distinction commonly made between this and ovoviviparity, in which fertilized eggs are retained within the mother, but with the embryo deriving nutriment solely from their yolk. However, it has been argued that the term is better avoided, since it is difficult to be sure that there is no other transfer of food between parent and young. On this view, all forms of retention of eggs and embryos should be termed viviparity.

In invertebrates

This mode of reproduction occurs sporadically in some invertebrates; in a few polychaete worms, for example, and in some periwinkles, but it is not always clear what its value is to the species concerned. Certain species of periwinkles that inhabit the upper shore illustrate the difficulty. Three species, *Littorina neglecta, L. patula* and *L. rudis*, are viviparous, and it would be reasonable to suppose that this is an adaptation to protect the eggs from desiccation when the tide is out, and to avoid having to discharge nauplius larvae into an elusive sea. But this makes it difficult to understand why *L. neritoides*, which lives in rocky crevices in the splash zone, and is only rarely immersed, is oviparous.

Viviparity amongst invertebrates is best known amongst insects. Apart from aphids, it is found also in cockroaches, which tend to live in dry habitats and which deposit their eggs protected within a tough capsule called the ootheca. Some species retain the ootheca within a brood pouch, and in certain of these there is a transfer of water and nutrients from parent to embryo, although it is not clear how the transfer is effected. Presumably the advantage gained is protection from desiccation, a factor which, as we shall see, has certainly been of importance in the evolution of the reproductive specializations of vertebrates.

Viviparity may involve adaptations for facilitating nutrient transfer, well exemplified amongst insects by the tsetse fly, *Glossina*, (the transmitter of the trypanosome that causes sleeping sickness in man). The eggs, produced one at a time, develop within a uterus (Fig. 9.11), into which a nutritive fluid is discharged through a nipple-like structure (a proceeding curiously reminiscent of the feeding of the young marsupial mammal in the maternal pouch). This provides for development up to the third instar larva, which then leaves the mother and pupates. Moulting within the uterus is

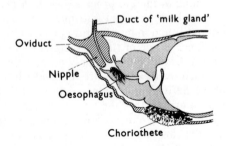

Fig. 9.11 Larva of tsetse fly *Glossina palpalis* in the uterus. (From Hogarth, P. J. (1976). *Viviparity.* Arnold, London.)

aided by an adhesive muscular pad, the choriothete, formed from the uterine wall; contraction of its muscles helps to strip away the moulted tissue. Some other viviparous insects emulate vertebrates still more by achieving a placenta-like structure, with close association of parental and embryonic tissues as a basis for nutrient transfer.

In vertebrates

Viviparity has been exploited to varying degrees in all the major groups of vertebrates, with the exception of birds, and with ecological advantages that are usually not difficult to evaluate. Outside the mammals, however, its occurrence is sporadic, which suggests that its benefits are not sufficiently marked in lower forms for viviparity to have a clear advantage.

It is found in some elasmobranch fish, where its evolution must have been encouraged by their practice of internal fertilization. Its advantage in this group may well lie in the inability of the embryo to regulate its urea level, which, as we have seen, is the basis of the osmoregulatory mechanism of this group. The simpler forms are essentially ovoviviparity, but in more advanced types the horny egg case (mermaid's purse) is reduced, and the yolk sac is applied to the oviduct wall, from which it absorbs nutriment.

Viviparity also occurs very sporadically in teleost fish, examples being the blenny of the sea shore (*Zoarces viviparus*), in which the egg develops within a central cavity in the ovary, and members of the family Poecilidae (familiar aquarium fish such as the guppy, *Poecilia reticulata*), in which development takes place within the egg follicle.

Viviparity also occurs in some amphibians. *Salamandra atra*, found in the Alps, gives birth to two young at a time, after a prolonged gestation, while a toad of tropical West Africa, *Nectophrynoides occidentalis*, carries its young in its oviduct, where they develop enlarged tails which are perhaps used as absorptive organs. Given the susceptibility of amphibians to desiccation, the advantage of viviparity is sufficiently obvious, and it may seem surprising that more of them have not become viviparous. The explanation is in part that some of them have adopted other and aberrant devices, such as carrying the young in their vocal sacs or in pits on their back. But mainly the reason is that the adults are themselves so sensitive to the need for water that most of them necessarily inhabit damp habitats where, in general, reliance upon laying eggs in the water, and fertilizing them there, has served them well.

Only in the higher vertebrates do we find a complex of integrated advances, of which reproductive devices are one major component, which has enabled these animals to leave the water completely at all stages of their life cycles, except when a secondary return has for some reason proved advantageous. The story goes back to the developmental adaptations which made it possible for reptiles to become fully established on land. Whereas the higher plants abandoned motile gametes, the vertebrates retained these, relying on internal fertilization, made possible by the development of appropriate mating behaviour and often, by the male, of a copulatory organ.The result, well established in reptiles and perfected in birds (Fig. 9.12), is that the oviduct can secrete around the egg a protective shell which provides a safeguard against desiccation, while permitting gaseous exchange with the atmosphere. Large amounts of reserve food (yolk) are deposited in a yolk sac, which is an extension of the embryonic alimentary tract, while the embryo is surrounded, within the shell, by folds (an outer chorion and an inner amnion) enclosing a cavity (amniotic cavity) containing water. This forms a protective cushion, substituting for the water in which the amphibian ancestors of the reptiles laid their eggs. Finally, the embryonic bladder enlarges to form the allantois, which, making close contact with the shell, provides for respiration and for the deposition of waste, while a layer of albumen (egg white) is also

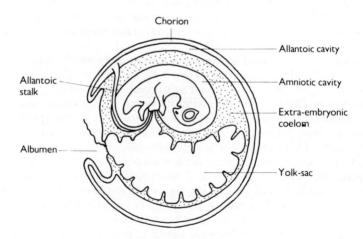

Fig. 9.12 Diagrams showing relations between extra-embryonic membranes and the chick embryo at the ninth day after incubation.

deposited; this layer, less conspicuous in lizards and snakes, provides a reserve of water for the growing embryo.

It has been suggested that these very remarkable adaptations were established in conditions of periodical drought and flood, in which those eggs that were laid higher up on the land would have been more likely to survive. But however this may be, the developmental advances made by the reptiles were of crucial importance for the evolution of birds and mammals. All three groups rely in one way or another upon the yolk sac, allantois, and amniotic membranes (collectively called the embryonic membranes), and for this reason are termed amniotes, contrasted with the anamniote fish and amphibians.

Some reptiles are ovoviviparous or viviparous, the egg shell being either reduced or lost. An example of ovoviviparity is the common lizard of Europe, *Lacerta viviparus*. In others, which are truly viviparous (the European skink, *Chalcides triactylus*, is an example), the yolk sac, in the absence of the egg shell, is closely applied to the oviducal (uterine) wall and absorbs secretion from it. Such an association of embryonic and maternal tissues for nutritive uptake, and for other forms of metabolic transfer, is termed a placenta. The chorion and allantois may also associate with the uterine wall to form a chorio-allantoic placenta, similar in principle to that of eutherians (p. 196), but of a relatively simple character.

Some form of live bearing has evolved at least 30 times recently in lizards and snakes, and many more species show egg retention, which is regarded as an intermediate evolutionary stage. The main benefit of viviparity in these animals is doubtless the protection of the young from environmental hazards during their earlier critical months. One plausible suggestion is that it is a thermoregulatory adaptation, correlated with cold climates and with maternal care of the young, although other possible ecological correlates come to mind, such as predation or extremes of soil moisture content. What is very clear, however, is that with the appearance at the reptilian stage of internal fertilization and the embryonic membranes, conditions were present favouring the evolution of viviparity when circumstances favoured it. Birds have never exploited it, doubtless because the weight of the retained eggs would have impeded flight. Moreover, the high level of instinctive behaviour achieved by these animals has encouraged the cooperation of both parents in guarding and feeding the young in well protected nests.

In mammals, however, viviparity has made a major contribution to their exploitation of terrestrial life, but it has done much more than that. With the capacity of mammals for flexible behaviour, both inborn and learned, viviparity is associated with patterns of social organization, in which one male may partner one female (this can, of course, happen in birds, too), or be associated with a group of these. This ensures food and protection, and also provides a basis for guiding the young, by learning and play, to gain essential experience before they have to fend for themselves. It need hardly be said that this has been the background to the evolution of our own species.

Three main division of mammals are recognized: the monotremes (e.g. platypus and spiny anteater), which are oviparous; the marsupials, in which the young are born in an exceedingly immature form and complete their development within the marsupium (pouch), taking milk from a nipple to which they become fixed; and the eutherians, comprising the main bulk of mammals, in which the young remain within the uterus until they have reached an advanced stage of development. It must be emphasized that these three groups are not successive stages in the evolution of viviparity in mammals; they are groups that have evolved along independent lines, marsupials being distinguished from eutherians by other features than just the state of the young at birth.

The female reproductive tract of marsupials is differently constructed from that of eutherians, for they substantially retain the paired tracts of the reptiles, with two uteri and lateral vaginae, in contrast to the fused uteri and median vagina of eutherians. The marsupial young must then be expelled through a median birth canal (temporary in a number of species, but permanent in others), and it may be this difference that has influenced the relatively short period of intrauterine development. To mention only one other difference, the basal metabolic rate of marsupials is some 30 % lower than that of eutherians, and, in correlation with this, their body temperature is lower, at 35.5° C, as compared with the 38° C typical of eutherians, and the 39.5–40.5° C of birds. This does not mean that thermoregulation is ineffective or imperfect in marsupials; we are only justified in saying that just as their reproductive anatomy is different, so their metabolic level is set lower. This could have advantages, such as lower food requirements, but it may also account for the intrauterine young developing more slowly than those of eutherians. However, this rate may also indicate that the transfer of nutrients from mother to foetus is less effective.

In eutherians this transfer has been promoted by the evolution of several types of chorio-allantoic placenta, formed, as in reptiles, by the close application of the chorion to the uterine wall, often followed by fusion and erosion of tissues. The resulting intimate association of maternal and embryonic tissue is then vascularized by the association with it, in lower mammals, of the allantois, carrying a blood supply. In higher mammals, including man, the mesoderm

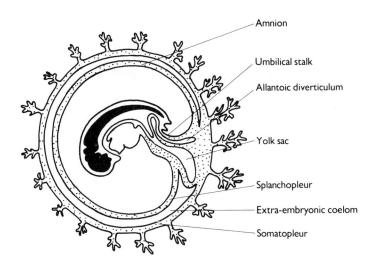

Amnion

Umbilical stalk

Allantoic diverticulum

Yolk sac

Splanchopleur

Extra-embryonic coelom

Somatopleur

Fig. 9.13   The stage of development of man at five weeks.

develops precociously and the allantois remains small (Fig. 9.13).
Amongst marsupials, only in the bandicoot (*Perameles*) is there a
true chorio-allantoic placenta, similar in principle of design to that
of eutherians, but different in detail and evolved quite
independently. In other marsupials the association of foetus and
mother is less intimate. The allantois remains small, and the
structure mainly concerned in nutrient transfer is the vascularized
yolk sac, which is only of minor importance in eutherians. The
association of maternal and foetal blood vessels in marsupials is
sufficiently close to provide for foetal respiration, and probably also
for some transfer of nutrients, but these are also taken up by the
absorption of uterine secretions into the yolk sac. The flexibility of
the reptilian solution to the problem of providing for development
out of water is sufficiently clear. It has been a major factor,
although by no means the only one, in bringing about the
dominating position of mammals in the conquest of the land.

10 Communication and Integration

Communication Systems

We have now seen ample evidence of the biological importance of communication and integration, defining the latter in general terms as the processing of signals so that living systems can give organized responses to them. They are needed for the life of the cell and for the orderly differentiation of the developing organism. They are needed to ensure the survival of individuals in fluctuating environments, for securing reproduction, and for maintaining the complex interspecific relationships that underly the organization of ecosystems.

An essential requirement for securing these results is the presence of communication systems within the body. These depend upon two types of pathway, conveniently distinguished as chemical and neural, although we shall see that there is a close functional and evolutionary relationship between them. Neural communication, found only in animals, is effected by nervous systems, based upon a highly specialized cell unit called the neuron. Chemical communication, found in both plants and animals, involves the transmission of chemical messengers by diffusion, as in plants and lower animals, and, in higher animals, by passage predominantly through the circulatory system. These messengers provide signals which evoke responses by interaction with receptors in what are called their target cells. Probably the origins of both types of system are to be found in the single-celled plants and animals. The activities of these are presumably integrated by chemical pathways, just as are those of the individual cells of multicellular organisms, but it may be that the fibre systems that can be identified in them represent some primitive form of neural transmission as well.

Phytohormones

The physiological processes of plants are largely, perhaps entirely, regulated by plant hormones (phytohormones). These are defined as regulators which are produced by the plant, are active in very low concentrations, and exert their effects at points separated from their sites of origin. The number of known compounds, both natural and synthetic, which act in this way is very large, and we can do no more here than give a few examples of them, and of their varied and interacting properties (Fig. 10.1).

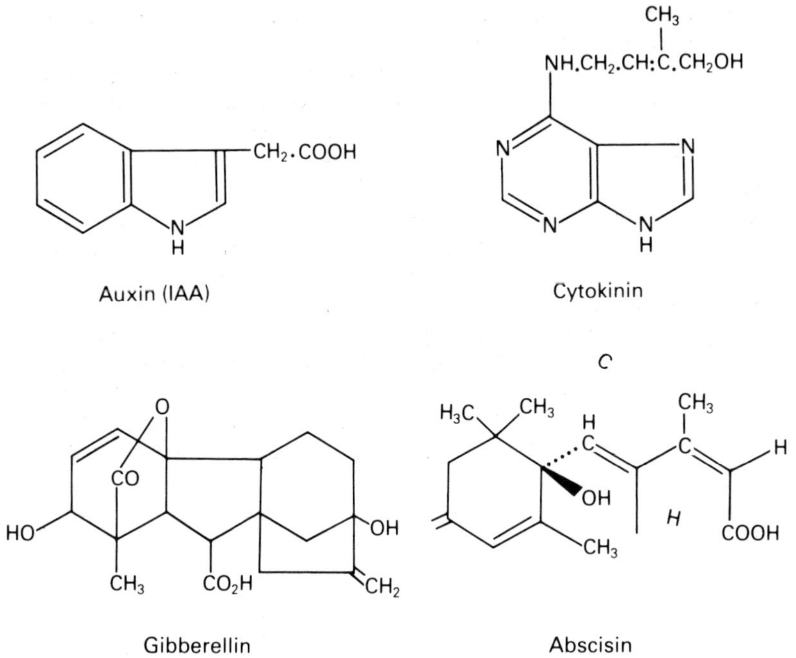

Fig. 10.1 Basic chemical structures of the major phytohormones.

One example is indole-3-acetic acid (IAA), a substance with many effects, but which is identified by its capacity to induce elongation of the cells of plant shoots. The generic term auxin is given to this and to related compounds (many of which are synthetic analogues) with similar actions. IAA was discovered during studies of the growth of the coleoptile of grass. During germination, the coleoptile, which is a sheath protecting the plumule, undergoes considerable elongation, which can be prevented by removing its tip. Growth can then be made to restart by placing on the tip a piece of agar containing material which has been allowed to enter it by diffusion from an excised tip. These and other experiments indicate the production of auxin in the tip, and its movement down the coleoptile to the elongating region. How the movement takes place is obscure, but it is not simple diffusion. It is clearly directional, dependent upon a polarity which is a property of the plant and which is often unaffected if the position of this is inverted. Moreover, the movement is not, in fully differentiated plants, restricted to the vascular bundles, but can take place through the parenchyma. There must be some form of specific transport mechanism, but the nature of this remains to be found.

Synthetic auxins, or compounds with auxin-like properties, have wide practical applications. Certain phenoxy compounds (e.g. 2, 4-d, 2, 4-dichlorophenoxyacetic acid) have auxin-like effects in low concentrations, but in higher ones can cause leaf deformation. They can thus be used to control weeds in lawns and pastures, while, deplorably, they have been exploited in warfare as a defoliant.

Auxin provides not only for unidirectional growth, but is also the means by which the growing plant is enabled to respond to unidirectional stimuli. These responses are called tropisms. Examples of them are phototropism (Fig. 10.2) and geotropism, which depend upon the movement of auxin being influenced by the stimulus exerted respectively by light or gravity. Experiments based upon collection of auxin from the coleoptile show that more of it passes to the side further from the source of light or nearer to the earth's centre. The effects of this are to promote asymmetrical growth in such a way that the coleoptile grows towards the light and against the force of gravity, i.e. upwards.

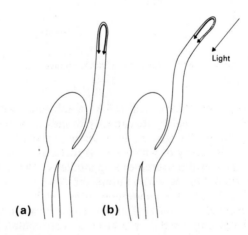

Fig. 10.2 Phototropism in a seedling. (**a**) Symmetrical distribution of auxin in the coleoptile in symmetrical light. (**b**) Unilateral light results in asymmetrical distribution of auxin, and consequent growth of coleoptile towards the source of light.

Such responses to environmental cues require receptors, which, because of the simpler morphology of plants, are much less obvious than in animals. The perception of gravity is thought to be a property of certain cells which contain starch granules. The position of these within the cells, which is influenced by the force of gravity, has effects upon the functioning of the cells. These effects are little understood, but it may be that they involve actions on cell membranes, or upon the distribution of organelles. In principle, however, the device is similar to the use by animals of various types of particles called statoliths, foreign or secreted, in their gravity receptors.

Photoreception, in plants as in animals, involves the use of pigments which both absorb and reflect light, although it is not easy to be sure exactly what pigment is being used. Carotenoids are likely candidates, but flavins have also been thought to be implicated. However, one pigment that is certainly involved in a wide range of plant responses to light is a pale blue material called phytochrome, widely distributed in all parts of plants, although in concentrations too low to be identifiable by its colour. Phytochrome (molecular weight about 60 000) is a protein conjugated with a pigment which is a tetrapyrrole (another illustration of the widespread importance of this type of compound). It exists in two forms, one (P_r) with an absorption peak at 660 nm (red) and the other (P_{fr}) with a peak at 730 nm (far-red). P_{fr}, which has been regarded as the active form, accumulates on exposure to sunlight, and is slowly converted to P_r in the dark. P_{fr} will inhibit flowering in short-day plants, while P_r inhibits it in long-day ones (i.e. those requiring a minimum number of hours for flowering). The existence of phytochrome accounts for so many aspects of plant photoperiodism being dependent upon red light.

Another illustration of phytohormone action in higher plants is given by the cytokinins (Fig. 10.1), substances which, in conjunction with auxins, promote cell division. An example is kinetin (6-furfuryl-aminopurine). In contrast to auxin, the cytokinins are mainly produced in roots, being then transported upwards through the xylem, although some downward movement through the phloem is also possible. Other functions reported for them include the promotion of lateral root formation, and the breaking of dormancy of buds. It is possible that the release of buds from apical dominance may depend on a balance between the concentrations of IAA and cytokinins. It is thought that auxins may influence the synthesis of RNA, and perhaps, through this, the production of enzymes involved in the synthesis of new cell material. There is evidence that the cytokinins, too, may influence enzyme production by acting upon RNA metabolism, more particularly in relation to tRNA. In both cases, however, elucidation of the chain of events is made difficult by these hormones having a wide spectrum of effects. What is clear, though, is the importance of balanced interaction in the functioning of these phytohormones, as in others that we shall mention.

Another group of hormones closely associated functionally with the auxins are the gibberellins (Fig. 10.1), which, in contrast to auxin, occur naturally in as many as thirty molecular variants. These substances, which are chemically related to terpenoids (e.g. steroids and carotenoids) take their name from their discovery in a fungus (*Gibberella fujikuroi*) which causes excessive growth of rice plants. Amongst their effects is the promoting of cell division, or elongation, or both, in stems, sometimes in cooperation with IAA but sometimes independently. Other functions include promotion of enzyme synthesis in germinating seeds, and involvement in the breaking of the dormancy of seeds and buds. Particularly striking is their ability to produce elongation in dwarf plants, so that these become indistinguishable from normal ones. Dwarfism is commonly due to a single mutant gene. This action of the gibberellins suggests that the effect of this gene may be to impede in some way the production of the hormone.

Certain phytohormones interact with those already mentioned by acting as growth inhibitors. One of these (Fig. 10.1) is abscisic acid (ABA), a terpene derivative which can act in amounts as small as 10^{-7}g. Synthesized in the leaf, it is passed to the shoot apex in the phloem, and sometimes also in the xylem. In addition to inhibiting cell division and elongation, it promotes leaf fall (abscission), apparently in antagonism to auxin, which retards the yellowing of leaves. Leaf fall has been thought to be associated with a fall in auxin levels in the leaf.

ABA is also involved, like auxin and the cytokinins, in apical dominance. Removal of the terminal bud, with a consequent release of lateral buds, is accompanied by a decrease in the ABA content of the latter, while application of ABA inhibits their development. Further, it promotes, at least in some species, seed dormancy. This form of dormancy (which, like that of buds, has obvious survival value) is not always due to inhibitors, but may simply be a consequence of the resistance of the seed coat. In the latter case, the dormancy will normally be broken by decay or rupture of the coat, but when inhibitors are the cause it is usual for a period of low temperature to be needed. The subsequent release is then associated with a rise in content of gibberellins. Bud dormancy, which is initiated in temperate zones in response to the shortening photoperiod of early autumn, and is induced by an increased abscisic acid content, prevents buds from opening during the winter, and is also correlated with increased frost resistance. As with dormant seeds, a period of chilling is often required before the dormancy can be broken; this breakage is associated with a fall in abscisic acid level and an increase in gibberellins.

Finally, ethylene, which is a hydrocarbon gas ($H_2C = CH_2$), and not, therefore, a substance of the type normally associated with hormonal action, is nevertheless regarded as a fruit-ripening hormone. Its output is greatly increased prior to the final stages of ripening, although how it acts is not clear. Treatment of fruit with ethylene is a well-known method of promoting their ripening, which can, alternatively, be delayed by removing ethylene from storage centres.

Nervous Systems and the Neuron

We have already mentioned that a fundamental difference between the communication systems of plants and animals is the existence in the latter of neural pathways. It would, indeed, be difficult to visualize animal life without them, for the motility of animals exposes them to a range of stimuli which far exceeds that encountered by plants, and creates a need for extreme speed and precision of response. A purely chemical mechanism, even with the advantage of a blood system, would be inadequate to provide for this. So it is that there is present, in all multicellular animals above the level of sponges, a nervous system formed of specialized cells called neurons. This system, in conjunction with receptor cells, provides for sensitivity to stimuli, and for converting (transducing) the information thereby gained into a form in which it can be transmitted through the body to those parts (effectors) which are to carry out appropriate responses.

A neuron (Fig. 10.3) is composed of a cell body (perikaryon) in which is contained the nucleus and various organelles, including mitochondria, Golgi bodies and endoplasmic reticulum. Particularly characteristic of it are cell processes, which are of two types, dendrites and axons. Dendrites are typically short and highly branched, while axons, of which there may be one or several, are long and usually branched only at their ends. It is along these processes that information is distributed, typically along structures called nerves, which are formed of numbers of axons.

In the simplest type of nervous system, exemplified in coelenterates, neurons are associated together as a nerve net, with individual neurons separated from each other by gaps (which are actually highly organized structures called synapses) between adjacent processes. Transmission in such a system takes place diffusely over the whole net, although there is often some tendency

204

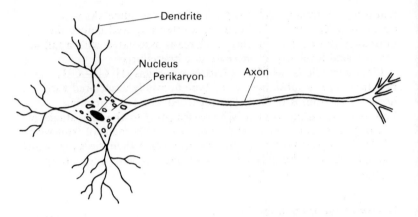

Fig. 10.3 A generalized neuron.

for fibres to become aggregated to form conduction tracts. This foreshadows the development found to varying degrees in almost all higher animals, in which neurons and their fibres, always separated by synapses, become concentrated in particular regions to form a central nervous system. This functions in part as an exchange, receiving and collating (integrating) information from many sources, and directing it to appropriate effectors. It can also generate spontaneous patterns of activity, so that movement, for example, may be initiated even in the absence of specific external stimuli. Further, it provides for storage of information, so that decisions can be taken in the light of past experience as well as with regard to new information received, the efficiency of adaptive response being thus greatly enhanced. Out of this emerges the fundamentally important capacity, possessed by different groups of animals to varying degrees, and expressed in different ways, called learning by experience. From this again there emerges, and particularly in ourselves, the capacity for insight learning, in which situations and problems can be seen as a whole, reflected upon, and resolved in new ways which were not apparent in the form in which the problem was initially presented. Scientific research is only one example of this capacity, as also is the reading of books written about it.

The transmission of information by neurons is essentially an electrochemical phenomenon (Fig. 10.4). Normally there is potential difference of some 70 mV across the cell membrane of the neuron, with the inner side negative. Transmission depends upon rapid and transitory changes in electric potential across the membrane, resulting from a brief change in its ionic permeability;

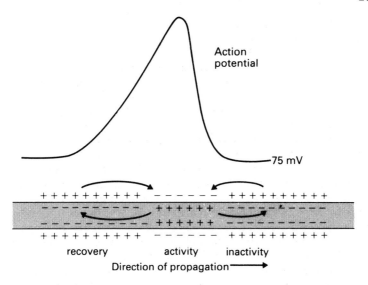

Action
potential

75 mV

recovery   activity   inactivity

Direction of propagation ⟶

Fig. 10.4 Diagram to show mechanisms of propagation of a nerve impulse. Arrows indicate local currents.

the effects of this include increased movement of sodium ions into the cell (lasting for about 0.5 msec), with the membrane potential becoming positive on the inside. This change of potential constitutes the action potential, the establishment of which is marked by a flow of electric current across the membrane. This flow is quickly terminated by a fall in sodium permeability and an increase in potassium permeability, which restores the resting potential, normally with a brief overshoot (for a few msec), in which the membrane potential differs slightly from the resting value. This is the after-potential.

The properties of the membrane and of the cytoplasm permit only a very restricted longitudinal flow of current down the axon from any one action potential. However, this potential creates a flow of current in the immediately adjacent area, and this results in depolarization of the membrane at that point, so that another action potential is produced. Continuation of this process down the length of the axon results in a wave of depolarization passing along it. We say, therefore, that action potentials are propagated along the fibre. The passage of the resultant wave constitutes the passage of a nerve impulse, and it is this impulse which is the unit of information transmitted through the nervous system.

One other detail is important in this brief review. Once an action potential has been generated, a brief period must elapse before the sequence of ion movements can be repeated. This is the refractory period, the duration of which limits the frequency of transmission of nerve impulses. Commonly this frequency does not exceed about 200 sec^{-1}. The consequence is that information is transmitted as a train of discrete impulses, and it is in this coded form that the diverse signals derived from the environment are distributed through the body.

Synapses may mark the boundary between an axon of one cell and the dendrite of another, or between an axon and the effector cell which it innervates. Synapses may be formed also at any region along an axon, or on the cell body, so that a diversity of pathways is available for the transmission and integration of information. Action potentials can be propagated in experimental conditions in either direction along an axon, but in the intact animal propagation is limited to one direction only, from the perikaryon and down the axon. This limitation is due to special properties of the synapse. Transmission across the synapse is sometimes by electrical means, the action potentials which arrive at the synapse generating potentials in the succeeding cell. More usually, however, transmission is chemical. It is then effected by the release at the synapse of a small amount of transmitter substance which produces the depolarization needed to generate a change of potential in the post-synaptic cell. One example of a chemical transmitter substance is acetylcholine, released at the junction of axons with vertebrate skeletal muscles, and at central synapses in these and other animals. It is almost immediately inactivated by an enzyme, cholinesterase, this ensuring that a single impulse arriving at the synapse will only generate a corresponding one in the postsynaptic cell. Other chemical transmitter substances include noradrenaline, which acts at the junction of axons with vertebrate smooth muscle, and also centrally, and 5-hydroxytryptamine, which functions especially in molluscs. Evidence is also accumulating of the involvement in this transmission function of a variety of peptides. Chemical transmission, which is widespread throughout the animal kingdom, shows that neurons are not only concerned with electrochemical phenomena, but are also secretory cells. This property, as we shall shortly see, has been of fundamental importance in the evolution of chemical modes of communication in animals.

Receptors

Animals have evolved a diversity of receptor cells, often grouped together with non-nervous cells into receptor organs. Two types of receptor cell are recognized. One is a modified neuron, connecting direct with the rest of the nervous system by its axon. This is called a primary sense cell. The other type is a non-nervous cell, synaptically connected with a neuron. This is a secondary sense cell. In either case the principle of action is that the sensory cell transduces the energy of the environmental stimulus into the electric energy of action potentials, which are then propagated in the manner already outlined. Continued stimulation results in continued excitation of the receptor, but each response is followed by a refractory period during which response is impossible. Stimulation thus generates a pulse of action potentials, and it is in this coded form that the information from all receptors is transmitted, regardless of the nature of the initial stimulus. The only variable in the code is the frequency with which the nerve impulses are propagated, for in general this frequency increases as the strength of the stimulus increases. The other important variable is that continued application of a stimulus usually leads eventually to the cessation of the nerve impulses. This is the phenomenon of adaptation. It may be slow adaptation, discharge of impulses being prolonged perhaps for a matter of hours, or fast adaptation, which may set in after one minute or less. Sensory adaptation secures economy of energy, for it enables animals to avoid responding to stimuli which have no significance for them.

Receptors can be usefully classified on the basis of the physical nature of the stimuli to which they respond, for usually, though not invariably, they are specialized in this regard. We can thus recognize chemoreceptors, responding to chemical stimuli; mechanoreceptors, responding to mechanical deformation, usually from low frequency mechanical stimulation; and electromagnetic receptors, responding to light and radiant heat over a spectrum that extends from the ultra-violet to the infra-red.

Chemoreception is perhaps the most primitive of the senses (a commonly used term, although we are only conscious of the sensation produced by our own receptors; the action of receptors in other animals can only be judged from their responses). Some chemical sensitivity must have been important from the earliest stages of evolution for the detection in water of food or of noxious material. This accounts for the virtually universal distribution of this capacity, and for its common association with primary sense cells. We customarily distinguish olfaction from taste, and the distinction has meaning. Olfaction, often associated with respiratory systems, is much the more sensitive and versatile, and is concerned particularly with stimuli coming from distant sources. Some olfactory receptors can be excited by a single molecule, and we shall see later the great ecological importance of this remarkable sensitivity. Taste is less sensitive, more limited in its range of response, and it deals primarily with nearby sources.

Chemical excitation takes place either at free nerve endings, or, more usually, on the cilia borne at the surface of the primary sense cells, but the mechanism is not understood. The major difficulty is to account for animals being able to discriminate by smell between large numbers of chemicals which are closely related to each other, and thus, amongst other things, being able to use olfaction for recognition not only of other members of the species, but also of individual members of it. Despite this, and here is another difficulty, molecules that are very different may smell the same. One possible explanation, amongst many, is that the reaction between molecule and receptor is a stereochemical one, depending on the presence in the cell membrane of specific sites onto which the reacting molecules can fit, while non-reacting ones have no sites available for them.

Mechanoreception is probably often provided for by free nerve endings, but sense organs with secondary sense cells also serve this function, especially in arthropods and vertebrates, which show some striking parallelisms in this regard. One widely exploited principle is the use of hair-like structures which are moved or deformed by pressure changes. An example is the use of sensory setae within the statocysts of many crustaceans (Fig. 10.5a). These organs are responsible for orientation of the body, the hairs being stimulated either by the pressure upon them of particles (statoliths) or by movements in the fluid brought about by changes in angular rotation.

There is here a close parallel with important sensory systems of vertebrates. The lateral-line system in the body wall of fish and a few amphibians contains groups of cells bearing sensory hairs which project into the gelatinous or fluid-filled cavity of canals (Fig. 10.5b). They provide a range of environmental information for the fish by responding to the low frequency vibrations that are set up by the flow of water, or by moving objects, or by reflection from stationary ones. The semicircular canals of the membranous labyrinth of the inner ear of vertebrates, thought to be derived from the lateral-line system, contain hair cells which similarly project into fluid. These provide information, derived from movements of the fluid, which aid the maintenance of equilibrium. Here are good examples of physical principles being exploited in similar adaptive specializations in unrelated organisms (see also p. 201).

The same principle of pressure stimulus acting upon hair cells has been exploited for auditory receptors, notably in the inner ear of vertebrates. The spectrum of sensitivity thereby achieved ranges in man from frequencies of 800 Hz to over 16 000 Hz, but birds

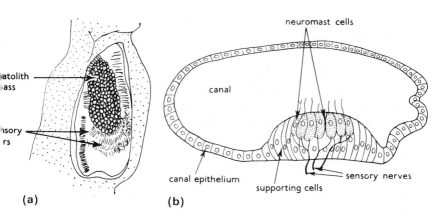

Fig. 10.5 (**a**) Diagram of the statocyst of the lobster *Homarus*, as seen on removal of the dorsal wall of the basal antennular segment. In this animal a number of small statoliths form a compact mass. (After Cohen (1955). *J. Physiol.*, **130**; from Wood, D. W. (1974). *Principles of Animal Physiology*, 2nd edition. Arnold, London.) (**b**) Transverse section of lateral line canal of the elasmobranch fish *Mustelus canis*. (After Johnson (1917). *J. comp. Neurol.*, **28**; from Wood, 1974.)

improve upon it, with a range up to 25 000 Hz. Even this, however, is exceeded by bats, which emit, and respond to, frequencies up to 100 kHz. This is the basis for their powers of echo-location, which enables them, by detecting reflection of these ultrasonic pulses, to avoid obstacles and pursue prey in the dark. Remarkably enough, however, certain insects, principally moths, react to this by themselves detecting these pulses and thus avoiding their predators; a powerful illustration of the resilience with which natural selection can secure the coordinated adaptation of members of an ecosystem.

It is not clear whether crustacean statocysts provide a sense of hearing, but insects certainly do have auditory organs, of high sensitivity. They present some analogy with those of terrestrial vertebrates, in that they depend upon vibration of membranous tympana, to which are attached sensory neurons. Thus are formed characteristic tympanal organs, found on various parts of the body, including the thorax and abdomen, and also the legs. The sensory neurons are arranged in groups, each with peak sensitivity to specific frequency bands. This provides for considerable auditory discrimination. Another factor contributing to this is the ability to use the exoskeleton for sound production, well exemplified in the male grasshopper, which produces sound by friction between its limbs and its tegmina (leathery forewings). These sounds serve as sexual recognition signals, and are an important element in intraspecific communication, many species having several different songs for different functions (e.g. territory holding, courtship, aggression towards the same sex, etc.).

Photosensitivity depends, as in plants, upon the absorption of radiant light by certain pigments, here called visual pigments. These are composed of carotenoids, conjugated with proteins. As we have seen, carotenoids are also present in systems associated with phototropic responses in plants, which suggests that these pigments were established early in evolution as agents in responses to radiant energy. There is some evidence, however, that flavins (and especially the widely distributed riboflavin) may be implicated in animals as they seem to be in plants.

Despite basic uniformity in the animal pigments (retinene, their carotenoid component, is very similar throughout the animal kingdom), there has been much evolution in the organs concerned. Photosensitivity is seen at its simplest, and presumably most primitive, in the dermal light sense, found in animals as diverse as amoeba and echinoderms; this does not use any structurally visible receptor system. Most animals, however, have developed also some form of eye, based upon cells (retinular cells) which contain photosensitive pigments. These pigments are activated by light in such a way that they bring about depolarization of the cell membrane and thus initiate the generation of nerve impulses. Sometimes, and especially in lower forms, the receptor cells are arranged in groups associated with other cells that contain a screening pigment (often the most conspicuous part of the eye, but not to be confused with the visual pigment), and often sunk in pits. Perhaps they sometimes permit the formation of simple images, but their main function is to respond to light intensity. Even this, however, can be the basis of essential adaptive responses, for on the information received, comparing the intensity of stimulation of the two sides of the body (these organs are typically paired in bilaterally symmetrical animals), the animal can direct its movement towards light or shade, as may be appropriate.

The most advanced types of eye are found in arthropods, cephalopod molluscs, and vertebrates. Those of arthropods (Fig. 10.6) are either simple (especially in young stages) or compound, the latter being so-called because they are composed of anatomical units called ommatidia. The association of photosensitive cells into ommatidia-like units is seen in certain polychaete worms, but the structures achieved in arthropods far surpass these, both in complexity of structure and of function. Apart from the expected discrimination of intensity, their unitary structure underlies the capacity of these animals to perceive the direction of incident light, and to detect movement in the surroundings. It is also the basis of their colour vision, which, as in vertebrates, depends upon the use

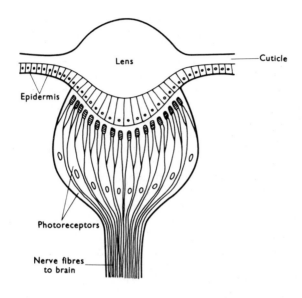

(a)

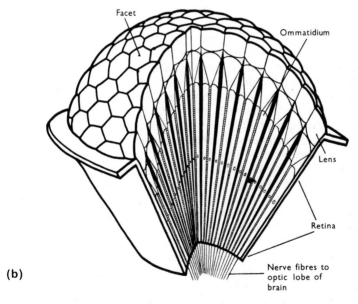

(b)

Fig. 10.6 Diagrams of invertebrate eyes. (**a**) Simple eye (ocellus) of a spider. The photoreceptors share a single cuticular lens. (**b**) Diagrammatic representation of the compound eye of an insect illustrating the arrangement of the ommatidia. (From Buchsbaum, 1976.)

of several kinds of pigment, each with a characteristic spectral sensitivity, and differentially distributed amongst the ommatidia. Further, arthropods are often sensitive to the plane of polarization of light. This remarkable range of visual capacities makes a major contribution to the success of the group, providing them with a correspondingly wide range of information, and making possible the organization of the complex lives of social insects such as bees, and the no less complex interactions between these and other insects with the reproductive functions of flowers.

The eye of vertebrates is fundamentally different in its organization, being constructed on the principle of a camera, in which respect it resembles the eye of cephalopod molluscs (octopods and squid), which is also able to form an image (Fig. 10.7). The vertebrate eye depends upon the focusing of an image (by the adjustable lens, and, in terrestrial vertebrates, by the cornea as well) upon a retina composed of two types of receptor cell, the rods and cones. These are present in very large numbers, up to 65 000 in the most sensitive region of the human retina, and it is this that gives to the vertebrate eye a power of resolution (that is, the capacity to discriminate between two adjacent points) very much greater than that of the compound eye. However, rods and cones differ in their resolving power. This is because each cone cell discharges information through a single nerve fibre, whereas a single fibre serves a large number of rod cells. Rods are more sensitive to light than are cones, and the effect of this is increased by the concentrated discharge from groups of rods. They are thus more plentiful in nocturnal animals, which are further aided by having at the base of their retina a reflecting layer (tapetum) which reflects light back onto the retinal cells after it has passed them. These devices are the basis of what is popularly known as 'seeing in the dark', their value counterbalancing the reduction of visual acuity which accompanies the convergence of the neural pathways of the rods. Conversely, diurnal birds, for example, may possess many cones but few rods, for they can afford the lower sensitivity, and profit by the enhanced acuity. In owls, rods may be entirely absent.

Colour vision is by no means so common in vertebrates as is sometimes inferred from its presence in ourselves. Especially characteristic of birds, it has been identified also in teleost fish, frogs, turtles and lizards, but it is present in only a few mammals (squirrels are one example), apart from man and other primates. It is generally supposed to depend upon the possession of three kinds of cone, differing in their pigments, and respectively sensitive to blue, green and red light.

Consideration of the plumage colours of birds, and of their sexual behaviour, shows the importance of colour vision in their

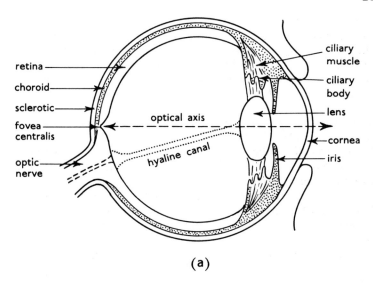

retina

choroid

sclerotic

fovea
centralis

optic
nerve

optical axis

hyaline canal

ciliary
muscle

ciliary
body

lens

cornea

iris

(a)

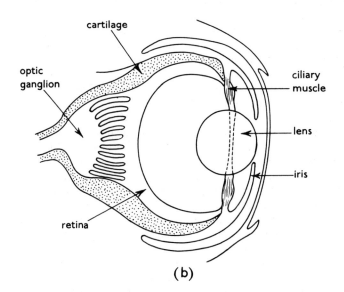

cartilage

optic
ganglion

retina

ciliary
muscle

lens

iris

(b)

Fig. 10.7 Diagram of sections through (**a**) a generalized vertebrate eye, and (**b**) a generalized cephalopod mollusc eye. The two chambers of the vertebrate eye are filled with a viscous material (anterior and posterior vitreous humour). (From Wood., D. W. (1974). *Principles of Animal Physiology*, 2nd edition. Arnold, London.)

intraspecific behaviour patterns, and it must also enhance their visual acuity during flight. It interacts also with the behaviour of other animals, as, for example, in the relationship of bird predators with their insect prey. Birds can learn to associate the colour of certain butterflies, for example, with their distastefulness, and are thus able to avoid them. This is exploited by insects in the formation of polymorphic (p. 19) mimetic groupings. In these an edible species may gain protection by mimicking a distasteful one (Batesian mimicry). This could be a disadvantage to the predator if the edible mimics outnumbered the distasteful models but natural selection ensures that this does not happen, by regulating the polymorphic balance. There are other circumstances, in which several distasteful species come to share a common colour pattern (Mullerian mimicry), in which the system works to the advantage of the predator by reducing the number of lessons which the young bird would otherwise have to learn.

The advantages of colour vision are not easy to define in more general terms. It is not easy to understand, for example, why in mammals it should be restricted, as far as is known, to so few species. Its absence from most diurnal mammals is correlated with the evident unimportance of colour in their sexual behaviour, as compared with birds and certain primates, but whether there is here a causal relationship is not clear.

One overriding consideration applies to all receptors. We have referred to their discriminatory power, but it is to be recalled that the information that they transmit is always a train of identical nerve impulses, the result of their transduction of environmental energy into action potentials. The interpretation of these impulses requires in ourselves a process of analysis which is the function of the central nervous system. The photochemical action of light upon visual receptors, for example, is transmitted to our cerebral visual centres. In principle, the same must be true of all animals, even in those with much simpler central nervous systems. Exactly what sensations they experience as a result of receptor stimulation we do not know. What we can see is that the receipt of information, in the form of stimuli, results in appropriate adaptive responses, because that information is transmitted, as nerve impulses, along pathways that direct them into the correct effectors. Those pathways are in part genetically determined, but they can also be built up or modified in the light of experience. These, of course, are easy words, stating the position with deceptive simplicity. The complexity of the processing that we are envisaging, and that underlies the maintenance of the environmental relationships of animals, can be judged by many criteria. One that is worth remembering is that the nose of an Alsatian dog, seeking out drugs or dead bodies, is estimated to contain some 300 million receptor units.

Animal Hormones

The nervous system of animals is complemented, in its functions of internal communication and regulation, by an endocrine system, responsible for the production of hormones. This is a chemical communication system analogous to that of plants, although of much greater complexity, in keeping with the different patterns of organization of plants and animals. The action of this system is now known to be more closely linked with the action of the nervous system than was earlier thought. The classical definition of hormones, derived from studies of higher vertebrates, is that they are secretions released directly into the blood stream from endocrine (internally secreting) glands, formed of epithelial cells. These hormones pass in the blood to other parts of the body, where, acting only in minute amounts, they evoke adaptive and regulatory actions. Such is the classical definition, but it has now to be applied in a flexible way, for two main reasons. One is that there are other chemical messengers which are essentially similar to hormones, except that they are distributed by diffusion instead of through the blood. This is naturally so in the lower animals (coelenterates, for example) which have not evolved a blood system, but which undoubtedly make use of chemical regulation. We have seen it to be true also of plants, in which also, of course, the hormones are not secreted by specialized endocrine glands.

The other reason is that some hormones are now known to be produced as peptide neurosecretions by specialized nerve cells called neurosecretory cells, and are thus not necessarily the product of epithelial endocrine glands. It is very likely, indeed, that this was the primitive mode of production of hormones, for it certainly predominates in the lower invertebrates, and it quite possibly evolved out of the production of peptide chemical transmitters to which we have just referred. Later, despite the evolution of epithelial endocrine glands, neurosecretory cells retained great importance because, as we shall see, they provide a link between the nervous and endocrine systems, and enable the latter to respond to environmental stimuli.

The actions of hormones extend to every aspect of function, from the regulation of growth and reproduction to the fine control of metabolic pathways. Some have already been mentioned (pp. 122, 144); one further example from vertebrates will illustrate the contribution that they make to metabolic regulation in that group.

Insulin is a hormone secreted in the pancreas by groups of cells located in the islets of Langerhans, which are distinct from the cells that secrete the pancreatic digestive enzymes. Its function is primarily the regulation of glucose levels in the blood. The release of insulin into the blood stream is stimulated by a rise in blood-sugar level, such as may occur during digestion. It then influences metabolic pathways in ways which contribute to lowering that level. Thus it promotes carbohydrate metabolism in muscle cells, and the storage in them of glycogen, largely by acting on the cell membrane to facilitate the active transport of glucose into the muscle cells. It acts upon the liver, not by influencing the cell membrane, for liver cells are always freely permeable to glucose, but by favouring a number of hepatic metabolic pathways, including glycogen storage. It acts also upon fat tissue, in somewhat the same way as on muscular tissue, and finally it promotes the uptake of amino acids in all three of these tissues, and the conversion of these acids into protein. Evidently the effects of insulin are complex and far-reaching, but they are not exerted in isolation. Just as in plants, physiological regulation is effected by the balanced action of a number of hormones, including, in this instance, secretions of the adrenal gland, the pituitary gland, and of other cells in the pancreatic islet tissue.

Neurosecretion

Emphasis is often placed upon an apparently sharp contrast between neural and endocrine coordination. The propagation of nerve impulses results in responses that are sharply localized in space and time, whereas hormones evoke responses that are widely distributed in the body, and that may be prolonged in their expression. Sometimes, too, hormonal responses are slow to develop, whereas the nervous system acts with immediate effect. The distinction is a real one, yet ultimately the endocrine glands must be regulated by the nervous system. This is because their action must be adaptively related to environmental signals; a requirement which becomes obvious when one reflects, for example, upon the way in which vertebrate reproductive cycles, regulated by hormones, are so often precisely related to changing photoperiod and other environmental cues.

This relationship is well seen in birds, where the breeding cycles are commonly adjusted so that optimum use can be made of seasonally variable food supplies. The adjustment is achieved through interaction of environmental stimuli with hormonal output. It is common, in temperate zones, for the gonads of birds to be of minimal size during winter, and for them to begin to enlarge when day-length increases in early spring. This enlargement involves not only the production of germ cells in the male, but an increasing output of male sex hormone from endocrine cells in the testis, promoting characteristic features of sexual behaviour such as song and the maintenance of territory. The female is slower in her sexual maturation, and requires, for completion of the process up to egg-laying, stimulation by the male in the mutual activities called courtship.

Photoperiod is a very common environmental cue in other vertebrates as well. In sheep, for example, the declining photoperiod of autumn evokes the start of the reproductive cycle, thus ensuring that the young are born at a favourable time of the year. Transfer of these animals from the northern to the southern hemisphere reverses the cycle, in terms of the calendar, so that the young continue to appear in the spring. All such relationships have, of course, been extensively exploited by man to modify the reproductive cycles of domestic stock to his own advantage. Battery hens are only one example.

There are circumstances, especially outside the temperate zone, where reliance upon length of daylight as a cue might be fatal. An example of this is provided by the weaver-finch, *Quelea quelea*, which is adapted for life in an arid climate. Its cue is rainfall, for this signals availability of nesting material and insect food, and the cue is so quickly followed up that nest-building, courtship and ovulation may be completed within eleven days.

It might be thought that the neural control of endocrine glands which is required by such responses could be achieved through the secretory cells being directly innervated, but this is not what is usually found. It is exceptional, certainly in vertebrates, for endocrine glands to have a secretomotor innervation; the signals to which they respond, therefore, can usually reach them only through the blood stream. How, then, can this arrangement provide for control of their secretory output? The answer is given by the neurosecretory cells. These contrast in certain features with conventional neurons, but, as we have already suggested, their existence within the nervous system is easier to understand when it is recalled that neurons themselves are engaged in secretion (p. 206) Neurosecretory cells resemble conventional neurons in having a cell body, with dendrites and an axon, but they tend to be larger, and to differ in their ultrastructure. The main difference, however, lies in their secretory product and its mode of discharge. This product is readily stainable, and is visible by light microscopy as material which arises in the cell body and passes down the axon to its ending. Electron microscopy shows that the secretion is contained in membrane-bound granules, of a size range of 100–300 nm, substantially larger than the vesicles in which chemical transmitters are contained. The neurosecretory axons often end in structures called neurohaemal organs, which are collections of neurosecretory nerve endings and blood vessels. Here the secretions are released and are passed into the blood stream, in which they circulate and function as hormones, just as though they had been produced in epithelial endocrine glands. Alternatively, however, they may be released for local action, without entering the main circulation, and even without entering the blood at all.

On these criteria, which are applicable both to vertebrates and to invertebrates (arthropods are so far the best studied of the latter), it is easy to distinguish between conventional neurons and neurosecretory cells, the fundamentally important property of the latter being that they secrete neurohormones. This property is a device for transforming neural signals into endocrine ones, and for the linking of endocrine regulation with neural regulation, by having certain hormones originate within the nervous system itself. We can illustrate this by brief reference to a very complex organ: the pituitary gland of vertebrates.

The pituitary gland (Fig. 10.8) is a dual organ, developing from two distinct sources. In part it is a neural organ, the neurohypophysis, which develops from the fore-brain as a down-growth of its infundibular cavity, and enlarges posteriorly into the neural lobe (pars nervosa). This lobe receives nerve fibres from the hypothalamus of the brain, where (in mammals) they originate in

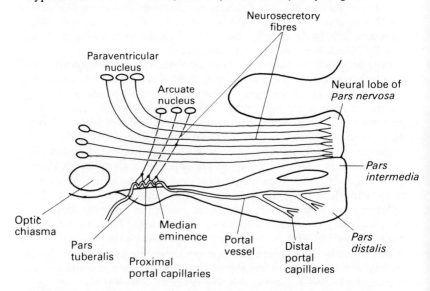

Fig. 10.8 Simplified diagram of the pituitary gland and associated structures in a mammal. Neurosecretions from groups of cells forming the paraventricular and supraoptic nuclei pass to the pars nervosa. Neurosecretions from groups of cells (e.g. arcuate nucleus) in the hypophysiotropic area of the hypothalamus pass to the median eminence.

two paired groups of neurosecretory cells. Neurosecretions pass down the axons of these cells into the neural lobe, from where they are released into the blood stream as two neurohypophysial hormones vasopressin (antidiuretic hormone) and oxytocin in mammals. Vasopressin is represented in lower forms by equivalent hormones differing slightly in molecular structure, and considerably in their properties; one of these is arginine vasotocin (p. 144). It follows that the neural lobe is a neurohaemal organ, for the release of neurohormones.

The other part of the pituitary gland, the adenohypophysis (pars tuberalis, pars intermedia, pars distalis), which develops from the buccal cavity, secretes a number of hormones, mainly in the pars distalis. They fall functionally into two groups. One of these, comprising growth hormone (somatotropin), prolactin, and two forms of melanocyte-stimulating hormone, consists of hormones which act directly upon their appropriate target tissues. The other group comprises hormones which regulate the activity of certain other endocrine glands, with which they are associated through a feed-back relationship (p. 20). Thyrotropin (TSH) regulates the thyroid gland, corticotropin (ACTH) regulates the adrenocortical tissue of the adrenal gland, and two gonadotropic hormones (FSH, follicle-stimulating hormone, and LH, luteinizing hormone) regulate the endocrine tissue of the gonads.

The actions of all of these hormones are certainly under the influence of the nervous system. Prolactin, for example, is released in response to the stimulus of suckling, and thyrotropin may be released in response to a fall in temperature of the surroundings, yet the only adenohypophysial cells that are directly innervated are the cells in the pars intermedia of the adenohypophysis which secrete the melanocyte-stimulating hormones that regulate the body colour of lower vertebrates by their action upon pigment-containing cells in the body wall. All the other hormones mentioned are secreted in the region of the adenohypophysis called the pars distalis, and none of these receives a nerve supply. How, then, are they brought under neural control?

The answer is that there is a second neurosecretory system associated with the pituitary gland, including further sets of cell bodies lying in the hypothalamus. Neurosecretions of these cells pass to a neurohaemal organ called the median eminence, which is a vascular area at the anterior end of the neurohypophysis (Fig. 10.8), containing neurosecretory nerve endings and the proximal capillaries of a vascular portal system (a term given to parts of a blood system which begin and end in capillaries). The neurosecretions are released into these capillaries, and are taken from there into the pars distalis by portal vessels which branch to form the distal capillaries of this system. The median eminence is thus the neurohaemal organ of a second neurosecretory system, entirely distinct from the neural lobe. The neurosecretions which traverse this system constitute a series of neurohormones which satisfy the classical definition in that they are transmitted to their targets through the blood stream, but are unusual in functioning close to their point of origin. They transmit signals from the hypothalamus, which has itself received signals from various receptors. The signals which thus reach the pars distalis specifically regulate the output of one or other of its hormones, either evoking their release or inhibiting it. According to which of these functions are served, the neurohormones are called releasing or inhibiting hormones. They, too, are part of the complex feed-back relationships which are fundamental to this pituitary system.

These, then, are the main pathways by which the central nervous system of mammals is able to regulate much of the endocrine system, and similar pathways operate in other vertebrate groups. They make possible the adjustment of hormonal balance in response to information which is derived from the receptors and led into the hypothalamus. They also provide for internal adjustments, since the output of those pituitary hormones which regulate target glands can be modified by feed-back from their targets. The output of thyrotropin, for example, is regulated by the level of thyroid hormone in the blood. The whole system is an excellent illustration of the role of feed-back in the maintenance of homeostasis in physiological systems (p. 65).

Brief reference to some invertebrate examples will serve to place neurosecretion in the wider perspective which it merits. There is convincing evidence for its involvement in the regulation of growth and reproduction in hydra (Fig. 10.9). Stainable droplets (seen, by electron microscopy, to be membrane-bound) appear in the nerve cells and fibres near the mouth (hypostomial region) and the bases of the tentacles during budding, and appear also in the same position in the buds themselves, just prior to their formation of their own tentacles. Further, asexual reproduction in this animal ceases at the onset of sexual reproduction, and, in correlation with this, the granules disappear. All of this suggests that the apical nerve cells secrete a neurosecretory growth-promoting hormone, and accords with long-standing experimental evidence that the hypostomial region exerts a regulatory influence on budding.

Well documented evidence comes from annelid worms, and especially from polychaetes. Sexually immature individuals of *Nereis diversicolor* (a ragworm) can regenerate their hind ends after their amputation (a valuable adaptation in creeping animals, vulnerable to bird predators). This power is lost if the brain is removed, but can

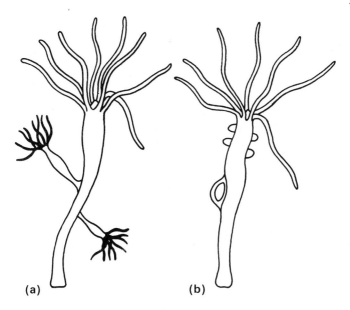

Fig. 10.9 Reproduction in *Hydra*. (**a**) Asexual, with two buds growing from the more basal (stalk) region. (**b**) Sexual, with several testes and one ovary (containing a single developing egg), developing in the more apical (gastric) region.

be restored by implanting the brain of another immature worm. This, in conjunction with much other evidence, indicates the production by the brain of a growth-promoting hormone. That this is a neurohormone seems certain, for its presence is associated with the existence of stainable granules in neurons of the brain, and in their axons. Just as in hydra, growth and sexual reproduction are mutually exclusive, a device which ensures economical application of energy. Removal of the brain from an immature worm results in it becoming precociously mature, which indicates the production of an inhibitory sex hormone by the brain of immature worms. It is likely that only one hormone is actually involved, the disappearance of the growth-promoting effect being accompanied by the lifting of the reproductive inhibition. Endocrine regulation, exemplifying similar principles, has been shown in other polychaetes, although there is much variation in the relationship of the controlling mechanism to the course of the life history.

Neurosecretory control certainly exists also in molluscs, but by far the best documented evidence comes from studies of arthropods, and especially of crustaceans and insects. It must suffice to say that in both of these two groups growth and moulting are under endocrine control, with neurosecretion playing a dominant part, although epithelial endocrine glands also contribute.

Neurohormones also regulate the colour changes which are found in many crustaceans, and which enable them to secure concealment by matching to the colour of their background. There is here a certain degree of analogy with colour control in the lower vertebrates (fish, amphibians, and certain reptiles), but in these groups it is either the conventional nervous system or the melanocyte-stimulating hormones of the pituitary that are concerned, or both, but not neurohormones. Many other functions are also hormonally controlled in crustaceans and insects, typically by neurohormones. Amongst them are water balance, protein synthesis, and the regulation of blood-sugar levels. Much remains to be learned about these aspects of endocrine regulation, in these and in other invertebrate groups, but the diversity already revealed in arthropods suggests a complexity matching that of the vertebrates.

Pheromones and Allelochemicals

We have so far been considering chemical regulation in terms of substances operating within a single individual, but so sensitive are organisms to chemical signals, and so varied are the possibilities of diversification of molecular structure, that chemical communication far transcends the limits of the individual, whether plant or animal. It has been possible, through the agency of natural selection, for highly specific, and therefore highly informative, relationships to be built up between individual species and their environment, both biotic and physical. It has been well said that the natural environment is a maze of chemical stimuli, often quite unappreciated by man, yet vital for the survival of plants and animals, and welding them into the ecosystems within which their lives are organized.

Chemical messengers operating in such ways between individuals, instead of within individuals, are of two types. One comprises those that are secreted by one individual of a species, and which evoke adaptive responses in other individuals of the same species. These substances are called pheromones. The second type comprises substances that pass from an individual of one species to an individual of another. No formal definition of this type has been formally adopted, but the terms 'allelochemical' and 'kairomone' have been suggested. There is here a large and complex field for investigation and review, but a few examples must suffice to illustrate the possibilities of these chemical signals. They can be found operating with impressive precision and elegance at all levels of organization.

We may suppose that the signals would initially have been of a generalized character, later evolving into much more precise and economical ones. Generalized allelochemical communication is well shown by the relationship between the blood fluke, *Schistosoma* (*Bilharzia*), and its invertebrate host. The fluke, which passes the early stages of its life cycle in certain species of aquatic snails, becomes in the later stages a major scourge of man. Motile cercaria larvae (p. 181) pass out of the snail and then enter humans who come into contact with the infected water. The parasite makes its way through the skin, with aid of histolytic secretions, and from there into the blood vessels. The eggs, which are laid in the vessels when the fluke has reached maturity, eventually reach the bladder or intestine and, where sanitation is poor, are passed out into the water where the snails are living. Miracidia larvae (p. 181) hatch from the eggs, and complete the life cycle by penetrating the body of the snail. Their choice of host is limited, only certain snail species are susceptible, and only certain strains within the species. Moreover, time is short, for the larvae must enter the snails within 24 hr if they are to survive. How, then, do they find their hosts?

In one sense the answer is that the parasite can only survive where the snail is abundant. This restricts its distribution, and, of course, it means that the infection can be attacked by eliminating the snail, should that be practicable. But there is more to the relationship than mere propinquity, for snails emit a chemical messenger which signals their presence to the miracidia larvae. The action of the messenger is shown by injecting suitable extracts of snail tissues into dishes of water in which the larvae are swimming. The effect upon the larvae is a clear-cut all-or-none response. As soon as they enter a zone in which the messenger is at a certain critical concentration, their movement becomes both more active and more restricted in range. When they reach the margin of the zone, they turn back into it, so that they tend to collect around the drop of extract. Under natural conditions, they would collect around the snails that are emitting the chemical. In addition, the messenger also arouses the penetration activities of the larvae. For example, they will attempt to burrow into pieces of washed snail skin or mucus if the water in which they are placed contains the substance, but they will not do so if the water does not contain it. Nor need the material that they attack be parts of the snails; they will even attempt to penetrate pieces of agar jelly which have been impregnated with the substance.

So the relationship between larva and host is promoted by a product of the host. The arrangement is clearly effective, yet it is not as specific as might be expected. Many snails, and not only the normal hosts, emit substances which activate the miracidia larvae of *Schistosoma*. The attractive substance is not, therefore, a specific one, nor, as we have seen, are the substrates into which the larvae burrow specific ones. The relationship, from this point of view, is a generalized one.

It might be thought that natural selection would act upon such a system to build up increasing specificity and perfection of action. That it has not done so in this instance, however, should not be regarded as a failure of evolution. Nature tends to economy. Where the end is adequately attained, there will be little pressure for improvement. Parasites survive primarily by producing prodigal numbers of eggs and offspring, only a few of which need to complete the life cycle. Vast numbers of miracidia are expendable. The chemical relationship between them and their snail populations aids their entry into the snail host, at the expense of the many larvae which are attracted to the wrong species, or which emerge into water that does not contain the required one. In this instance, therefore, chemical messengers are merely one element in the total ecological situation.

The freshwater parasitic copepod, *Lernaeopoda edwardsii*, gives an example of greater precision. This animal lives on the gills of brook trout, but not on those of the rainbow trout or the German brown trout. This restricted distribution is determined by a specific chemical which diffuses into the water from the gills of the brook trout, and which activates the larvae of the parasite, as a result of which they establish themselves on the gills of this species. The other two species do not produce this secretion, so that the larvae show no response at all to their gills, even when they come into direct contact with them. Thus, if individuals of the three species of fish are all exposed to the larvae, it is only the brook trout which become infected.

These have been examples of allelochemicals. Pheromones have been particularly richly demonstrated in insects (Fig. 10.10), partly because it was in this group that their importance was first clearly demonstrated, and also because the high level of specialization of these animals is reflected in the remarkable ways in which these substances, and also allelochemicals, are exploited. To take an example almost at random, the Large White butterfly (*Pieris brassicae*), which flutters irritatingly around our cabbages, lays fewer eggs on leaves that already carry eggs or larvae, or that are damaged. Visual cues are probably one factor influencing its behaviour, but so also is a pheromone associated with the eggs; present, perhaps, in the cement or deposited by the previous female from her accessory glands.

Especially well known are the pheromones used in the colonial life of the honey bee. Not the least remarkable feature of them is their multiplicity of actions, and the economy of secretory organization thereby secured. One of the pheromones is 9-oxodecenoic acid (*trans*-9-keto-2-decenoic acid). This is secreted by the queen, licked

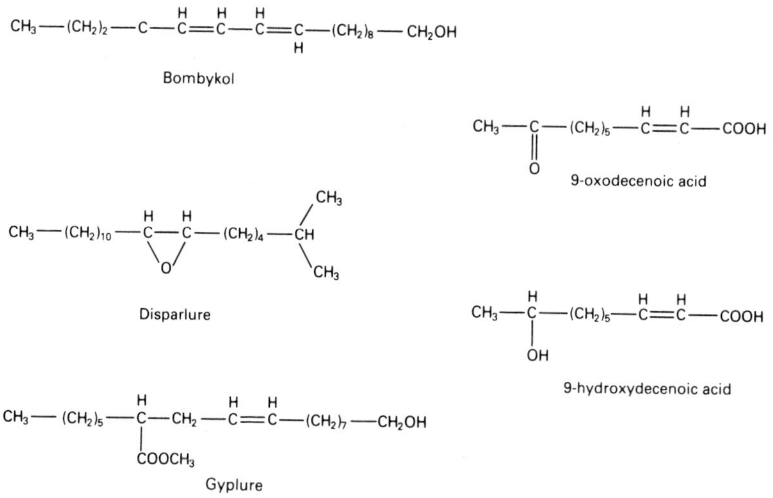

CH_3—$(CH_2)_2$—C—$\overset{H}{\underset{}{C}}$=$\overset{H}{\underset{H}{C}}$—$\overset{H}{\underset{}{C}}$=C—$(CH_2)_8$—$CH_2OH$

Bombykol

CH_3—$\overset{}{\underset{\overset{\|}{O}}{C}}$—$(CH_2)_5$—$\overset{H}{\underset{}{C}}$=$\overset{H}{\underset{}{C}}$—COOH

9-oxodecenoic acid

CH_3—$(CH_2)_{10}$—$\overset{H}{\underset{}{C}}$—$\overset{H}{\underset{O}{C}}$—$(CH_2)_4$—$\overset{}{\underset{}{CH}}\underset{CH_3}{\overset{CH_3}{}}$

Disparlure

CH_3—$\overset{H}{\underset{OH}{C}}$—$(CH_2)_5$—$\overset{H}{\underset{}{C}}$=$\overset{H}{\underset{}{C}}$—COOH

9-hydroxydecenoic acid

CH_3—$(CH_2)_5$—$\overset{H}{\underset{COOCH_3}{C}}$—$CH_2$—$\overset{H}{\underset{}{C}}$=$\overset{H}{\underset{}{C}}$—$(CH_2)_7$—$CH_2OH$

Gyplure

Fig. 10.10 Pheromones of insects.

from her body by her attendant workers, and spread by them through the colony as they regurgitate their food. In conjunction with a second and volatile pheromone, 9-hydroxydecenoic acid (*trans*-9-hydroxy-decenoic acid), also secreted by the queen, it inhibits the development of ovaries in the workers and the construction by them of queen cells. 9-oxodeceonic acid is also a sex attractant, drawing drones to the queen during the nuptial flight, and it acts, too, as an aphrodisiac, stimulating the drone to mount the queen. Its rate of output by the queen falls off towards the end of her life, thus permitting the rearing of new queens. Both pheromones also function during swarming. 9-hydroxodecenoic acid activates the restless bees, and 9-hydroxydecenoic acid quietens them after they have settled; another contribution to economy of energy.

Pheromones are extensively used by female moths as sex attractants, and many of these substances have now been chemically characterized. Bombykol, secreted by the female silkmoth, is one example, and disparlure, secreted by the female gypsy moth (*Porthetria dispar*), another (Fig. 10.10). These substances are remarkably potent, or, to put the matter in another way, the males' receptors are exceedingly sensitive. Thus gyplure, another substance secreted by the gypsy moth, and thought at first to be the sex attractant, can lure males to traps in the field that contain only 10^{-7} mg. Male moths may also secrete pheromones; these seem to act as aphrodisiacs when the two sexes have been brought together.

Of great ecological interest are the complex interactions that have developed between different types of molecular messenger. They are well exemplified by the polyphemus moth (*Antherea polyphemus*), the larvae of which will feed only on the leaves of the red-oak. Attention was drawn to this species when it was observed that females in outdoor cages attracted males over large distances, whereas those reared and maintained in the laboratory consistently failed to mate successfully, even though males were present with them. The explanation is that the female releases a sex pheromone which acts on the antennae of the male, and which is needed to attract it and to activate it sexually, much as with *Bombyx*. This pheromone, however, can only be produced by the female in response to a volatile *trans*-2-hexanal which is produced by the oak leaves, and which acts on antennal receptors of the female. A remarkable consequence of this is that females from which the antennae have been severed can attract males for up to six hours after the operation. This is because nerve impulses are propagated from the injured ends of the antennae, and provide an adequate stimulus for release of the pheromones, comparable to the impulses which would normally be generated by the plant chemical. An important element of this chemical inter-relationship between tree and insect is that many other plants also produce the *trans*-2-hexanal, but they also release masking substances which prevent the female from responding. Altogether, then, we find here an elegantly adaptive interaction between a plant allelochemical and an animal pheromone, which ensures that reproduction and oviposition shall take place only in the presence of the very leaves which are obligatory food for the larvae.

Considering that some moths are major pests (the caterpillar of the gypsy moth is a serious pest of trees), and that the control of these and other insects by man-made toxins is proving increasingly unacceptable to thoughtful opinion, the use of sex pheromones for ecological control has great attractions. One might hope that insects would not react to these substances, as they have to chemical insecticides, by developing resistant strains such as have contributed so seriously to the difficulties of eradicating malaria, for example. More than twenty species of the mosquitoes which transmit the malaria parasite have developed resistance to organochlorine insecticides, while one species is reported to have responded by changing its behaviour, resting out of doors after biting its human victim and thus avoiding the sprays used in the house. But there are still many difficulties to be overcome in the use of pheromones. The chemical aspects are complex, for the amounts of secretion available for investigation may be only obtainable in nanogram quantities. The problem is increased by the sex pheromones of moths being often mixtures of two or more substances, an arrangement which doubtless

favours the securing of specificity. An example is given by a
defoliating pest, the spruce budworm. A sex pheromone,
characterized as *trans*-11-tetradecanal, was obtained from some
200 000 female moths (*Choristoneura fumiferana*), but only after
extensive field trials was it found that a small percentage (4 %) of the
cis compound had to be present as well to achieve attraction.

Ecological difficulties are also many. Factors such as the spacing
of trees, in the case of forest pests, the area to be covered, and
fluctuating weather conditions, may make trapping impracticable
under natural conditions, although it may be feasible in the more
uniform conditions obtaining during horticultural production. But
should trapping not be possible, there remain other approaches, such
as distributing sex pheromones so as to disrupt communication and
thus prevent the males from finding the females. It remains to be
shown how useful this type of method may prove to be on a
commercial scale. It is most likely to be of value as a sensitive
indicator of population size, that may be used to determine whether
or not control measures will be needed.

We may conclude with a few examples from vertebrates.
Freshwater minnows (*Phoxinus*), and other freshwater fish belonging
to the large group Ostariophysi, signal the presence of danger to
other members of their species by the release of alarm substances
from their skin. The reaction is highly specific, for the messenger is a
pheromone, produced only in certain cells of the skin. It is only
discharged when this is damaged, and is not released from fish that
are already dead, nor from tissues other than skin. Little is known of
the chemistry, but the potency of the messengers is shown by a
solution prepared from 0.1 g of skin in 100 ml of water inducing a
response in minnows when it is added to a 150 l aquarium. The
response is mediated through olfactory receptors, and is not,
therefore, shown by minnows in which the olfactory nerve has been
cut and the olfactory bulb of the brain removed. It will be observed
that even in an aquatic habitat the pheromone is an olfactory one.

The extraordinary level of specialization and sensitivity achieved
by chemical communication is well illustrated by the salmon. Eggs
are laid in fresh water, the alevins yielding parr which later
metamorphose into smolt, these migrating down rivers into the sea
and maturing there into grilse or salmon. Abundant experimental
evidence shows that these fish, despite travelling great distances at
sea, can yet find their way back to the rivers in which they began
their lives. For this they use chemical cues, including the products of
vegetation which characterize the home water, and which must have
been imprinted as a memory early in their lives. This is shown by
presenting mature salmon with the choice of two currents of water,
one containing extracts of their home water, or of its vegetation, and

the other lacking these. The results of such experiments demonstrate both their memory and their powers of discrimination, for they are able to recognize the water in which they were reared. Further evidence comes by recording spontaneous electric potentials from their olfactory bulbs, while infusing their nasal cavities with home water or with water from other sources. It is the home water that decisively enhances the electrical activity.

It is the very advanced level of neural and sensory organization of these fish which enable them to extract the maximum of information from their chemical environment. A moment's reflection will show, however, that this can be only part of the story, for it is not conceivable that they can recognize the qualities of their home river while they are still far out at sea. Other cues are needed. These are not yet fully understood, but are thought to include the ability to navigate by the sun, and perhaps also by the moon. It would take us too far afield to pursue the matter further, but at least we are reminded that chemical messengers are only one element of the extraordinarily complex relationships which have been built up between animals and their environment.

The operation of pheromones in mammals is seen in the laboratory mouse, the reproductive (oestrous) cycles of which are influenced by a volatile pheromone (the Whitten effect). This is shown by placing females in a wind tunnel with males, but separated from them. Those which are 2 m downwind of the males show a greater proportion of individuals in oestrus than do a similar group of females which are 2 m upwind of the males. The biological significance of this in the normal life of mice is not clear, although it is possible that females might be attracted to males by odour, and that some effect upon their oestrous cycle might be an adaptively beneficial consequence of this.

Closer to human experience are the apes and old-world monkeys, which have a menstrual cycle, one well-studied example being the Rhesus monkey. Sexual behaviour in the male is aroused by chemical messengers present in the vaginal secretions of the female, and capable of stimulating the olfactory receptors of her partner. Her sex attractant odours are derived from a mixture of fatty acids in the vaginal secretion, their production depending upon the action upon the vaginal tissues of the female sex hormone (oestrogen). It is antagonized by another sex hormone, progesterone; this is released at ovulation, so that its inhibitory action upon the vaginal tissues ensures that attractiveness to the male will be maximal at ovulation. Remarkably, these fatty acids (acetic, propionic, *iso*-butyric and *iso*-valeric acids) are not produced directly by the female, but by bacteria in the vagina. This somewhat bizarre example of symbiosis (p. 54) provides experimenters with three means of protecting the female monkey from the attentions of the male. Ovariectomy is one, progesterone treatment is another, while treatment with antibiotics is a third and perhaps the simplest. It will be noted that here again we have an example of an interaction between hormones and pheromones.

Whether vaginal secretions are of any physiological importance as messengers in human sexual activities, or whether they may have been so in the past, is not clear, although the matter is open to experimental and observational testing. A curious illustration comes from studies of 135 female students who were living together in a hostel of a women's college in the United States. Women who met males less than three times a week were compared with those who met them three or more times a week, excluding from consideration those who were taking the contraceptive pill. The two groups showed a statistically significant difference in the length of their menstrual cycles. Those with less frequent exposure to males had a cycle of 30.0 ± 3.9 days, while those with the more frequent exposure had a cycle of 28.15 ± 2.9 days, which is close to the normal length. Perhaps there is a suggestion here of some physiological factor influencing the reproductive cycle, interestingly reminiscent of the Whitten effect just noted. But if so, its nature is obscure, if indeed it exists, and it certainly cannot be concluded, from these data alone, that the cycles of women are influenced, as are those of mice, by an olfactory pheromone from the male. Nevertheless, it is of cognate interest that women are much better able than are men to perceive certain musk-like substances, including boar taint substance, which is produced by the preputial glands of the boar, and which is said to be identifiable also in men.

The problem of possible sex pheromones in ourselves is complicated by the reduction of our olfactory sensitivity, in comparison with other mammals, and by the importance of psychic factors, including the feeling that natural odours are unaesthetic. On the other hand, the profitability of the perfume industry, which also exploits animal musks, suggests that artifical sex attractants are not without their importance in human environmental biology. The use of artificial sex repellants, however, which has been whimsically suggested as a possible contraceptive aid, may well prove less appealing.

11 Outline Classification of Living Organisms

Biological classification is intended to provide a hierarchical system in which the members of any one group are related by common descent. Such a system constitutes a natural classification, in contrast to artificial classifications which are based on superficial features (whales grouped with fish, for example), and which take no account of evolutionary relationships. The general principles of the classification of living organisms are not in dispute, but there are differences in detail between one system and another, and points of disagreement remain, some of which have been indicated earlier in this book. Examples are the nature of viruses, the status of bacteria and fungi in relation to green plants, and the interrelationships of plant-like and animal-like unicellular organisms with each other and with multicellular ones.

Systems of classification will be found in many text books. The one which follows is a simplified but workable outline, to be used in close conjunction with the main text and the index. It should help readers to place the groups within an evolutionary framework, and to relate them to each other, where this is possible. It should be noted that in the classification of the plant kingdom the term *Division* is preferred to the term *Phylum*, which is used in animal classification.

Further reference may be made to:

BELL, P. R. and WOODCOCK, C. F. L. (1971). *The Diversity of Green Plants,* 2nd edition. Arnold, London.

CAIN, A. J. (1974). Biological classification. In *Encyclopaedia Britannica, Macropaedia* vol. **4**.

KINGDOM PROKARYOTA (bacteria and blue-green algae (Cyanophyta), and perhaps viruses)

KINGDOM FUNGI (MYCOTA)

KINGDOM PROTISTA (unicellular and multicellular algae, and Protozoa; the Fungi are sometimes included here)
SUBKINGDOM PROTOPHYTA (algae)
SUBKINGDOM PROTOZOA (unicellular animals; these reappear below as

a subphylum of the animal kingdom, the difference in treatment depending primarily upon a botanical or zoological approach)

KINGDOM PLANTAE (METAPHYTA)
    Division Bryophyta (no differentiated vascular system)
        Class Hepaticae (liverworts)
        Class Musci (mosses)
    Division Tracheophyta (with a differentiated vascular system)
        Subdivision Psilopsida (e.g. *Psilotum*)
        Subdivision Lycopsida (e.g. *Lycopodium*, club moss)
        Subdivision Sphenopsida (e.g. *Equisetum*, the horsetail)
        Subdivision Pteropsida
            Class Filicinae (ferns)
            Class Gymnospermae
            Class Angiospermae
                Subclass Monocotyledonae
                Subclass Dicotyledonae

KINGDOM ANIMALIA
SUBKINGDOM PROTOZOA (unicellular animals, with related plant-like forms; see also above)
SUBKINGDOM PARAZOA
    Phylum Porifera (sponges; multicellular animals differing from all others (metazoans) in having a less integrated organization, with no nervous system)
SUBKINGDOM METAZOA (all other multicellular animals)
    Phylum Cnidaria (often called coelenterates; with tissues, but lacking organs)
    Phylum Ctenophora (similar in certain respects to cnidarians, but perhaps not closely related to them)
    Phylum Platyhelminthes (flatworms, with no blood system)
    Phylum Nemertinea (more advanced flatworms, with a blood system)
    Phylum Aschelminthes (round worms, including nematodes)
    Phylum Rotifera (wheel animalcules)
    Phylum Annelida (segmented worms)
        Class Polychaeta (marine bristle worms)
        Class Oligochaeta (including earthworms)
        Class Hirudinea (leeches)
    Phylum Sipuncula
    Phylum Pogonophora
    Phylum Arthropoda (with a hard and jointed exoskeleton; by far the largest phylum in number of species)
        Class Crustacea (copepods, shrimps, crabs, etc.)
        Class Onychophora (e.g. *Peripatus*)
        Class Myriapoda (centipedes, millipedes, etc.)

    Class Insecta
    Class Merostomata (including king-crabs)
    Class Arachnida (scorpions, spiders, mites, etc.)
Phylum Mollusca
    Class Monoplacophora
    Class Polyplacophora (chitons)
    Class Aplacophora
    Class Gastropoda (limpets, sea slugs, snails, etc.)
    Class Scaphopoda (tusk shells)
    Class Bivalvia (e.g. mussel, oyster)
    Class Cephalopoda (e.g. squid, octopus)
Phylum Echinodermata (sea stars, sea urchins, sea cucumbers, etc.)
Phylum Hemichordata (related to echinoderms and to chordates)
Phylum Chordata (with a supporting notochord at some stage of
the life cycle)
    Subphylum Urochordata (= Tunicata)
    Subphylum Cephalochordata (e.g. *Branchiostoma*, = amphioxus)
    Subphylum Vertebrata (with a cranium and vertebral column)
      Superclass Agnatha (without jaws)
        Class Cyclostomata (lampreys, hagfish)
        (and several extinct classes)
      Superclass Gnathostomata (with jaws)
        Class Placodermi (extinct fish)
        Class Elasmobranchii (cartilaginous sharks and rays)
        Class Actinopterygii (bony fish, including teleosts)
        Class Crossopterygii (bony fish, including lung fishes)
        Class Amphibia
        Class Reptilia
        Class Aves
        Class Mammalia

Further Reading

AMBROSE, E. J. and EASTY, D. M. (1977). *The Cell*, 2nd edition. Nelson, London.

BARCROFT, J. (1938). *Features in the Architecture of Physiological Function*. Cambridge University Press, London.

BARNES, R. D. (1974). *Invertebrate Zoology*, 3rd edition. Saunders, Philadelphia and London.

BARRINGTON, E. J. W. (1968). *The Chemical Basis of Physiological Regulation*. Scott, Foresman, Glenview, Illinois.

BARRINGTON, E. J. W. (1975). *An Introduction to General and Comparative Endocrinology*, 2nd edition. Clarendon Press, Oxford.

BARRINGTON, E. J. W. (1979). *Invertebrate Structure and Function*, 2nd edition. Nelson, London.

BELL, P. R. and WOODCOCK, C. F. L. (1971). *The Diversity of Green Plants*, 2nd edition. Arnold, London.

BIRCH, M. C. (ed.) (1974). *Pheromones*. North Holland, Amsterdam.

BOLD, H. C. (1977). *The Plant Kingdom*, 4th edition. Prentice Hall, Englewood Cliffs, New Jersey.

BUCHSBAUM, R. (1976). *Animals without Backbones*, revised edition. University Press, Chicago.

CHAPMAN, G. (1967). *The Body Fluids and their Functions*. Studies in Biology, no. 8. Arnold, London.

CHAPMAN, R. F. (1971). *The Insects*, 2nd edition. Hodder and Stoughton, London.

COLINVAUX, P. A. (1973). *Introduction to Ecology*. Wiley, New York.

DIXON, A. F. G. (1973). *The Biology of Aphids*. Studies in Biology, no. 44. Arnold, London.

DYSON, R. D. (1978). *Cell Biology: a Molecular Approach*, 2nd edition. Allyn and Bacon, London.

GORDON, M. S. (1979). *Animal Physiology: Principles and Adaptations*, 3rd edition. Macmillan, New York.

HAGGIS, G. H. (1974). *Introduction to Molecular Biology*, 2nd edition. Longmans, London.

HARDY, R. N. (1972). *Temperature and Animal Life, 2nd edition*. Studies in Biology, no. 35. Arnold, London.

HOPKINS, C. H. (1978). *Structure and Function of Cells*. Saunders, Philadelphia and London.

IMMS, A. D. (1973). *Insect Natural History*. Fontana New Naturalist, Collins, London.

JAMES, W. O. (1973). *An Introduction to Plant Physiology*, 7th edition. University Press, Oxford.

JENNINGS, J. B. (1972). *Feeding, Digestion and Assimilation in Animals*, 2nd edition. Macmillan, London

MACAN, T. T. (1974). *Freshwater Ecology*, 2nd edition. Longmans, London.

MAITLAND, P. S. (1978). *Biology of Freshwaters*. Blackie, Glasgow and London.

MANNING, A. (1979). *An Introduction to Animal Behaviour*, 3rd edition. Arnold, London.

MEADOWS, P. S. and CAMPBELL, J. I. (1978). *An Introduction to Marine Science*. Blackie, Glasgow and London.

MEGLITSCH, P. A. (1972). *Invertebrate Zoology*, 2nd edition. University Press, Oxford.

MILL, P. J. (1972). *Respiration in the Invertebrates*. Macmillan, London.

MORRIS, J. G. (1974). *A Biologist's Physical Chemistry*, 2nd edition. Arnold, London.

ODUM, E. P. (1966). *Ecology*. Holt, Rinehart and Winston, New York.

PHILLIPSON, J. (1966). *Ecological Energetics*. Studies in Biology, no. 1. Arnold, London.

POSTGATE, J. (1978). *Nitrogen Fixation*. Studies in Biology, no. 92. Arnold, London.

POTTS, W. T. W. and PARRY, G. (1964). *Osmotic and Ionic Regulation in Animals*. Pergamon, Oxford.

PROSSER, C. L. (ed.) (1973). *Comparative Animal Physiology*, 3rd edition. Saunders, Philadelphia and London.

RAY, P. M. (1972). *The Living Plant*, 2nd edition. Holt, Rinehart and Winston, New York.

ROSE, S. (1976). *The Chemistry of Life*. Penguin Books.

ROUTH, J. I. (1978). *Introduction to Biochemistry*, 2nd edition. Saunders, Philadelphia and London.

SCHMIDT-NIELSEN, K. (1964). *Desert Animals*. Clarendon Press, Oxford.

SCHMIDT-NIELSEN, K. (1979). *Animal Physiology: Adaptation and Environment*, 2nd edition. University Press, Cambridge.

SMYTH, J. D. (1976), *Introduction to Animal Parasitology*, 2nd edition. Hodder and Stoughton, London.

STEVENSON, G. B. (1970). *The Biology of Fungi, Bacteria and Viruses* 2nd edition. Arnold, London.

STREET, H. E. and OPIK, H. (1976). *The Physiology of Flowering Plants*, 2nd edition. Arnold, London.

SUTCLIFFE, J. (1978). *Plants and Temperature*. Studies in Biology, no. 86. Arnold, London.

SUTTIE, J. W. (1977). *Introduction to Biochemistry*, 2nd edition. Holt, Rinehart and Winston, New York.

SWANSON, C. P. and WEBSTER, P. L. (1977). *The Cell*, 4th edition. Prentice Hall, Englewood Cliffs, Jersey.

USHERWOOD, P. N. R. (1973). *Nervous Systems*. Studies in Biology, no. 36. Arnold, London.

VAN DIBBEN, W. H. and LOWE-McCONNELL, R. H. (eds) (1975). *Unifying Concepts in Ecology*. W. Junk B. V., The Hague.

Various authors (1978). Evolution. *Scientific American*, **239** (3).

WHITFIELD, P. J. (1979). *The Biology of Parasitism*. Arnold, London.

WHITTAKER, T. H. (1975). *Communities and Ecosystems*. Macmillan, London.

WIGGLESWORTH, V. B. (1974). *Insect Physiology*, 7th edition. Science Paper Books, Chapman and Hall, London.

YOUNG, J. Z. (1962). *The Life of Vertebrates*, 2nd edition. Clarendon Press, Oxford.

Index